Information Transducers

V.M. Bogomol'nyi

///VSP///

ISBN: 90 6764 426 9

PRINTED IN THE NETHERLANDS BY RIDDERPRINT BV, RIDDERKERK
COVER DESIGN: ALEXANDER SILBERSTEIN

Contents

Introduction

The US Department of Commerce has defined the most promising directions in the design of new materials and technologies: *micropositioning* and monitoring of material fatigue, life expectancy and *investment lifetime, sensors* and *actuators* (in scanning tunnel microscopes), *piezoelectrics* and *composites.*

Piezoelectronic devices are widely used in the USA and Japan. In 1995 US industry manufactured 1,400 million dollar worth of piezoceramic elements (the selling price of one piezoceramic element is $20). In 2002 the American output was two billion dollars. US Government allocates about four billion dollars annually for the research in this field [1].

Piezoactive crystal, amorphous and polymer materials are used in electronics, automatic devices, computers and measurement devices. This book considers piezoelectric transducers based on ceramic polar dielectrics (DE) and semiconductors, which are most often applied in information systems [1-3, 6, 9, 11, 12, 18, 23, 24, 33, 37, 40-44, 61, 65, 66, 69, 74, 85, 86].

The direct piezoelectric effect (the electrization of polar dielectrics under mechanical stresses) occurs in the non-symmetrical crystal lattice (mostly as a result of ion and orientation polarization). Polar dielectrics under an alternating electric field develop a *"biasing current"* (i.e. the capacity current of condenser charging and discharging). It is a specific current without free charges on the electrodes of metal-dielectric-metal (MDM) structures.

All biological sensors are DE. They include protein crystals, dipoles of molecules (e.g. OH⁻ hydroxyls), as well as the most complex molecular structures, life-supporting biological membranes. Human and animal tissues have *piezoelectric characteristics.* There are about 2,400 solid, liquid and gaseous DE (organic and non-organic substances) in nature, which greatly exceeds the number of known metals and semiconductors (about 60-80). The principal feature of polar DE is that they can retain electrical fields up to 10^5 *V/cm* for hundreds of years (the figure is given for precious stones). In metals the charge is lost within 10^{-15} s, whereas the electric field strength cannot exceed 0,001 V/cm.

Polar DE used in biosensors react only to *changes in the electric signal.* Thus, *redundant information is eliminated* in the simplest way. It is difficult to imagine how the brain would function if thousands of millions signals from cells entered it at the same time.

The unusual and varied properties of ferroelectrics (FE) account for their wide application in science and technology. Metallic conductivity and polarization in dielectrics work together to produce the effect of high-temperature "superconductivity". The nature of this phenomenon is not fully known. However, thin ferroelectric ceramic films show an abnormally high conductivity at room temperature.

Piezoelectric crystals (niter as gunpowder, salts of tartaric acid) were known as early as in the 16th century. In 1756, Aepinus found opposite charges on the two electroded surfaces of a heated tourmaline crystal. In 1790, Rene-Just Hauy, the French crystallographer, studied the pyroelectric effect (the appearance of static electricity produced by a change of temperature over time). In 1880, Pierre and Jacques Curie began a quantitative study of quartz, tourmaline and Rochelle salt crystals. *The first piezoelectric measurement device, the quartz balance, was used by Pierre and Marie Curie in 1901 for preparing milligrams of radium emanation.*

In the beginning of the previous century, Robert Wood and William Cady demonstrated the unusual possibilities of quartz as a source of ultrasound waves (USW) and an autogenerator for electronic time devices. K.Shilovsky and Paul Langevin established that quartz plates under an alternating electric field radiate USW in water for dozens and hundreds of kilometers and proposed to use them in hydroacoustic devices. Industrial application of piezoelectric transducers begins in 1914 with K.Shilovsky and Langevin's invention of the sonar, which was used during World War I. Initially, crystals of quartz and Rochelle salt were used as a source and receiver of oscillations. Piezomodules of Rochelle salt are dozens of times as big as analogous constants for piezoceramics PZT-4, which has the widest industrial application [1, 11, 12].

William Cady discovered the autosynchronizing effect and showed that quartz stabilizes external electric fields at resonance. In 1920 he built quartz frequency stabilizers and microphones [2]. Valašek studied the unusual properties of Rochelle salt: unipolarity, hysteresis, the Curie point (phase transfer temperature) [2]. Valašek created piezoelectric sound receivers, which were used in artillery reconnaissance. Contemporary acoustic and radio electronic devices allow one to identify working aeroplane engines at a range of thousands of kilometers and listen to conversations inside buildings hundreds of metres away (300 m).

Ferroelectrics (nonlinear piezoelectrics) include pyroelectric materials used in night vision devices. At night in reflected starlight one can "see" just like in the daytime [1, 61, 65, 66, 88].

Valašek discovered the hysteresis characteristic of polar dielectrics. Nonlinear properties of piezo- and ferroelectrics (FE) are currently used in various manufactured solid-state electronics (varistors, nonlinear varicaps controlled by the electric field, tandels, i.e. ferroelectric devices maintaining constant temperature). The energy dissipated by tandels under switching spontaneous polarization is used to stabilize operating temperature at the Curie point.

Electroconductivity of polar DE was first considered in the doctoral theses of Jacques Curie (1899) and A.F.Ioffe (1904), W.C.Röntgen's postgraduate. Since then Academician A.F.Ioffe and his team studied the electric strength and electroconductivity of DE.

B.M.Vul and I.M.Goldman (1945) discovered ferroelectric properties in barium titanate ($BaTiO_3$) crystals. Almost at the same time similar research was carried out in the USA and Japan. In 1952, the Russian scientists worked out a *technology for industrial manufacture of ceramic polar dielectrics* (polycrystal solid solutions of metal oxides): lead titanate zirconate (PZT-4, PZT-5), commonly used in industry [1, 8, 9, 11, 12].

Industrial manufacture of piezoceramic materials and their unique and universal properties account for their use in different fields of technology: mechanical engineering, electronics and computers. These features have been used in fundamental scientific research (superconductivity) [77].

At present there are 1-60 μm thick composite materials and ferroelectric ceramic films. Thin 1-2 micron films are used as sensors and computer memory elements. The unique electrophysical properties of *piezoceramic materials* are superior to the strength characteristics of monocrystals [9, 11, 12, 23, 37, 38].

An analog electric circuit, in which inductivity L, capacity C and resistance R are connected in series, can be used as an integral characteristic of the electrophysical properties of the piezoelement. Capacity C_0, which characterizes the high ohmical resistance of the piezoelement and depends on its geometry, is included parallel to the L-C-R chain. This layout explains the universal properties of piezoceramic elements [1, 2, 5, 6].

Piezoceramic elements can be manufactured in any shape and size, which is their advantage over monocrystals. High *dynamic strength*, comparable to the *strength of steel* (fatigue limit under cyclic load $\sigma_{-1} = 30$ MPa), functionality at the frequencies of 100 MHz and higher and temperatures from -50^0 C to $+ 600^0$ C, as well as high chemical and radiation durability account for DE's industrial application [1, 2, 5].

Piezopolymers are also used in industry [3, 24]. In 1995 the "elastic" transistor, which consists only of polymers (a seksitophen semiconductor, a polyether insulator, graphite-based polymer electrodes), was built. Electret microphones with a 200 μm fluoroplastic film are manufactured in all the industrial powers of the world (with the annual output of about 200 million).

Calculation of a thin-film piezoelectric pressure sensor from oriented PVDF

Material	Relative dielectric permeability	Piezoelectric coefficients		Pyroelectric coefficient
	ε_{33}^{T}	d, pK/N	G, V*cm/N	$\left[\dfrac{nC}{cm^2\,grad} \right]$
BaTiO$_3$	1700	190	1.3	20
Polyvinylidene chloride	3.5	0.7	2.3	0.1
Nylon – 11	3.7	0.26	0.8	0.5
PVDF (Polyvinylidene fluoride)	12	28	26.4	4

Table 1. Comparison of pyro- and piezoelectric properties of barium titanate and piezopolymers.

When the film is under compression by force F, there is an electrical field strength E_0

$$E_0 = \frac{36\pi Fdh}{\varepsilon_{33}^{T} S \cdot 10^{-9}},$$

where d is the piezomodule, $\varepsilon_{33}^{T} = 11$ is the relative dielectric permeability, F/S=1.0 [N/m^2], S is the area of the sensor, S=10^{-4} m^2. The stated values for PVDF give E_0=0.06 [V/m].

"Piezoelectric material deforms by a very small amount when an electric field is applied. For example, when 100 V is applied across a 9 μm-thick PVF$_2$ film, the dimensional change in the polymer chain direction is 3 μm for each centimeter of length. The motion amplitude magnification ratio of a bimorph is roughly the ratio of its length to the thickness. Since 6-7 μm is the approximate minimum thickness of a PVF$_2$ film, a magnification of a few thousand times is easily obtained.

"The maximum applicable field without degrading piezoelectric parameters is 300 kV/cm for PVF$_2$, and 3 kV/cm for PZT. For this reason the displacement of a PVF$_2$ bimorph can be very large. For example, the displacement of the free end of a 2-cm long, two-layer bimorph cantilever structure is 1 cm when 100 V is applied across its 9 μm-thick films.

"The PVF$_2$ bimorph is suitable for driving very light weight structures; for example, for light control, for rf or dc power control, or for air flow generation.

"In the case of composite piezoelectric polymer material in which PZT powder is bound by a PVF$_2$ matrix, the bimorph structure was investigated for loudspeakers and telephone-type microphones" [M.Toda. Voltage-induced large amplitude bending device – PVF$_2$ dimorph – its properties and applications. *Ferroelectrics*. 1981. Vol.32, p.127-133].

The composite, a polymer matrix filled with piezoceramic grains (optimal ceramics ratio is 30%), is characterized by a sensibility ten times greater than *monolithic piezoceramics*. Japanese scientists have devised a tactile polymer sensor for robots, the sensibility of which can be compared with that of a human finger [6], [Ma X., Brett P.N. The performance of 1-D distributive tactile sensing system for detecting the position, weight and width of contacting load. *IEEE Trans. on Instrumentation and Measurements*. 2002. Vol.51. No.2, p. 331-336].

Powerful hydroacoustic sonars are used in household appliances such as washing machines, dry-cleaning machines, as well as in USW devices for intensifying technological processes in food industry (sterilization of dairy products), pharmaceutical industry, metallurgy (for manufacturing ultrastrong aluminum alloys). Sonars are constructed from piezoceramic ring plates connected with a metal bolt. The bolt should be pre-drawn very carefully to eliminate any chance of brittle fracture. For this one can use special contact compression sensors [Silva D.G. et al. Strain gauge tactile sensor for finger-mounted applications. *IEEE Trans. on Instrumentation and Measurements*. 2002. Vol. 51, No.1, p.18-22].

Highly elastic piezopolymer materials (including resin-based) have better mechanical performance characteristics as compared to *strong, but brittle* piezoceramics, which performs poorly under *bending, tension and shock load* [9, 11-13].

Piezoelectric chemical resonance sensors have an unusually high sensibility. A 1 *nanogram* change in the external mechanical load causes a *1 Hz resonance frequency drift* (a thin piezoelectric plate is covered by a sensitive thin layer, which adsorbs the chemically controlled substance).

In practice, direct piezoeffect sensors of position, pressure and acceleration (accelerometers) are most widely used. Stress and strain sensors measure stress ranging from 1 to $2*10^5$ N with an accuracy of 1% and relative strains up to 10^{-5} [1, 17, 21, 23, 24, 74].

Vibroacoustic sensors (noise and vibration sensors, fluid level sensors) are used for the non-destructive testing (NDT) of machines and equipment, as well as in vehicle-borne and stationary diagnostic systems in automobiles [1, 37].

Piezoelectric actuators create displacements with *the discrete spacing of one angstrom to fractions of millimeter* (the displacement is unlimited). They are used in devices measuring the size of machine parts and surface roughness, miniature relays, devices with a response speed of less than 1 μs, focusing devices in optical systems (shutters with a mechanical force of about 1 MN, spacing up to 0.1 mm, and 20 μs response speed). The accuracy of positioning (a discrete spacing of 0.01 μm) and speed of piezomotors ensure precise autotracking in studio sound and video recording equipment and micropositioners for fiber-optic systems [18, 24, 29, 65, 66].

Piezoelectric sources of displacements (actuators) are used in scanning tunnel and atom-force microscopes. Graphic display piezoelectric motors develop a speed of 700-800 mm/s. Piezoelectric linear displacement motors are used for recording holographic information in board-to-surface systems [1, 9]. These motors have a discrete spacing of 0.01 μm, displacement force of 6 N, and speed of 200 mm/s.

In the process of recording, the scanning laser beam moves over discrete points on the holographic plate and records information in three dimensions. An 8 x 9 cm plate made of optically transparent material can hold 10 thousand documents. The moving laser beam smears the image; on the other hand, the beam in the contemporary construction is stationary and the holographic plate with 0.01 μm spacing is moved by the piezomotor [8, 18]. In the latter case the dynamic image becomes static.

Piezoelectric actuators are used in electric field-directed filters with tunable frequency, dielectric resonators, and transducers with 10% tuning range and millisecond speed [1, 2, 5, 8].

Among piezoelectronic devices with *double electromechanical transformation* (a combination of the direct and inverse piezoeffects) the most widely used are resonators, filters (accounting for more than a half of the industrial output of piezoceramic elements), piezotransformers of voltage, current and wattage. These devices are used, for example, in *TV line timebase units as* 38-120 MHz (6.5 MHz) frequency transducers, filters, resonators (37 kHz) and time-delay lines. *In electronic watches* piezoceramics and quartz generate *oscillations* at 37 kHz [9, 11].

Apart from conventional bipolar resonators, there are multielectrode ones, with *an acoustic connection* between "in" and "out". They improve electric control of resonator parameters, eliminate undesirable oscillation modes and create band-pass filters with greater performance characteristics [37].

Due to the high *(1 Hz per nanogram)* sensibility of their resonance frequency towards external influence, piezoceramic elements are used in "secondary" sensors for measuring mechanical, heat and electrophysical data, as well as the chemistry of fluids [24, 38].

Computer interface makes use of combinations of various physical phenomena, as follows from contemporary developments: the infrared optical mouse, the USW plotting board digitizer with a USW "pencil" which reads graphical information.

A sensor for controlling *frost formation in the freezing chamber* is an example of how various physical effects work together. This device is based on an optoelectronic element. When frost precipitates on the condenser, the monitored surface reflects less light, and the change is registered by the photodetector. The layout of the optoelectronic sensor is shown in Fig.1. The light-emitting diode is made of alloyed silicon carbide, gallium arsenide (GaAs) or polycrystal piezosemiconductor. Light-emitting diode 1 radiates at wavelength $\lambda = 0.85\text{-}1.3$ μm. The light, reflected by the monitored surface, is received by photodetector 2, whereupon the photocurrent enters amplifier 3.

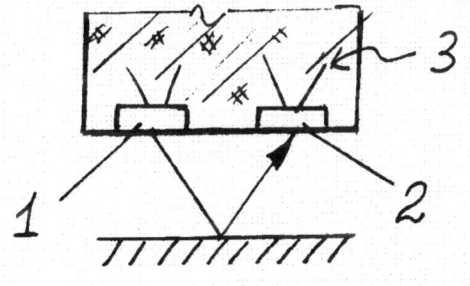

Fig. 1.

In the aircraft frost-up control system frost thickness on the condenser wall is measured by the resonance frequency drift of the piezoceramic resonance sensor.

Remote sensing of surface (cracks and other structural defects) is performed with the help of charge-coupled devices (CCD). An object is "lit" with an electric lamp (its infrared radiation can be transmitted over fiberglass), the reflected light is received by CCD, which send the information to the computer.

A CCD inside a miniature robot, which crawls on the inner surface of a 110 mm tube (the cooling system in atomic reactors), is able to sense cracks and other defects on the inner surface of the tube at the distance of 200 meters. Image detail is controlled by the piezomotor, which moves the lenses of the optical system.

Solid-state (monolithic) structures and piezoelectronic *integral and functional* devices have been recently implemented in single-chip computers. These devices make use of nonlinear properties of ferroelectrics [1, 9].

The functional approach to the manufacturing of information systems is based not on the *combination of multiple elements* in the integrated circuit, but on the integrated *use of the physical properties* of the crystal. The required functional properties are achieved with specific-shape electrodes and controlling signals.

At present there is a wide variety of information systems:
- memory elements (bimorph, one- and multi-layer planar matrices; hybrid memory modules);
- analog elements (summing units, multiplication-division devices, code-electric strength transducers);
- multifunctional adaptive analog-digital devices (multiplexers, digitizers);
- combined functional media: ferroelectric-semiconductor, solid-state integral matrices, tactile sensors (computer keyboard);
- CPU clock generator;
- piezomotors of disk drives and printing devices;
- electromechanical interface devices (keyboard, mouse, audio source, ink-jet printers)

The possibility of electron transport in DE was shown by Mott and Gurney in 1940; in 1952 it was confirmed experimentally. MDM structures, i.e. CdS-based *dielectric diodes*, which work *in the mode of space-charge-limited currents* [3, 108], appeared in the early 1970 s.

High ohmical piezosemiconductors (semiinsulators) include gallium arsenide (GaAs) with the width of the "prohibited" zone $\Delta E = 1.43$ eV, which is characteristic of DE with $\Delta E = (1 \div 5)$ eV. GaAs-based switching diodes have a reactive speed of less than 0.5 ns. GaAs crystals are sources of coherent radiation with the wavelength $\lambda = 0.6\text{-}0.9$ μm, as well as fast-response IR and visible light receivers. Gallium arsenide may be used for direct transformation of solar energy into electric energy (efficiency ~17.6%).

Electric field microwave (MW) oscillations, found in GaAs monocrystals, were used to manufacture simple miniature generators of MW electromagnetic coherent waves (a 500 MHz device for remote assessment of car speed). This miniature radar has the capacity of 5-10 W in the non-stop mode and 200 W in the impulse mode.

Industry manufactures a solid-state *counterpart of the electronic vacuum lamp*, the *submicron* permeable base field transistor. This GaAs-GaN crystal-based construction has a "net", which is embedded between the cathode and anode by X-ray lithography. The energy consumption of the transistor (1.2 μm thick) is about one femtowatt. Its amplification is 16 dB (the standard field transistor works at 5-10 mW).

The dielectric transistor with a semiconductor layer or metallic net as the controlling element has the following merits:

- low noise,
- high input resistance,
- relatively low temperature sensibility.
- Research into emission currents in DE is of practical interest due to the following reasons:
- vital importance of information about *leakage* through insulating substrates (used in microelectronics) and their temperature behaviour,
- manufacture of computer memory devices and nonsilver photosensitive films used in electrophotography.

These issues are studied by *dielectric electronics*. Traditionally, active materials are divided into *dielectrics* and *semiconductors*. Thus, there are two branches of solid-state physics: *semiconductor* and *dielectric physics* [A.A.Potapov Molecular dielcometry (in Russian). Novosibirsk: Nauka. 1994].

Semiconductor physics deals with: transport, generation and recombination of free charge carriers (electrons, ions and "holes"), as well as electrophysical phenomena in *metal-semiconductor contacts* (Schottky barriers). *Dielectric physics* studies the polarization and electric strength of crystals and nonlinear behaviour of MDM structures: condensers, piezotransducers of voltage and current, delay lines, dielectric waveguides and resonators. *Dielectric electronics* studies *emission currents in DE; injection and recombination of free charge carriers in DE* [3, 7].

Piezoelectronic integrated devices represent *monolithic ceramic circuits* (MCC), manufactured from combinations of polar dielectrics and semiconductors. They can be space and planar (2.5 to 30-50 μm) [1].

Devices based on surface acoustic waves (SAW) are used for signal processing: delay lines, filters, phase revolvers, generators (at frequencies higher than 1 MHz) [8, 18].

Monolithic ceramic circuits (MCC) are more technological in comparison with other constructions of microcircuits, while being considerably *cheaper* to produce. The density of packaging equals $(5 \div 10)*10^3$ elements/cm^2 (3,5∗10^3 elements/cm^2 for monolithic semiconductor circuits) [9, 11, 12, 18].

Information processing systems comprise piezoelectric acoustic-optical devices (AOD) for processing radio signals, light and sound waves in real time. AOD are used for modulating light, transforming optic radiation into electric signals, and analyzing radio signal spectrums [Filatova E.Yu. Optimizing transmission of acoustic-optical cell with sectioned piezotransducer. Ph. dissertation thesis. M.V.Lomonosov Moscow State University: 2002; Parygin V.N. et al. Improvement of the acousto-optic cell function by piezotransducer sectioning. *Journ. of Modern Optics.* 2000. Vol.47. No.9, p.1501-1509; Filatova E.Yu., Parygin V.N. Transmission function of acoustic-optical cell with apodized piezotransducer. *Journ. of Pure and Appl. Optics.* 2001. Vol.3. No.4, p. 540-545].

Tunable acoustic-optical filters (AOF) receive electromagnetic radiation and extract a narrow-range component, the center band of which is controlled by the acoustic signal (with a narrow pass band about several angstroms). AOF are used for controlling tunable lasers, spectral analysis of images, channel separation in optical communication lines, compression of light impulses and detection of impurities in gases.

The pass function, i.e. the dependence of diffracted light intensity on its wavelength (with a given sound frequency), is one of the main characteristics of AOF. Apart from the basic maximum, the pass function develops "lateral lobes" (from 5 to 12%). This decreases the dynamic range of AOF (the ratio of the maximum signal received by AOF to the minimum signal).

To change the pass function of the acoustic-optical cell, one has to change the distribution of the sound field amplitude along the light beam. The simplest way to an uneven distribution of the sound amplitude is to dissect the piezotransducer into several parts of different length and apply different voltages to each section (Fig. 2).

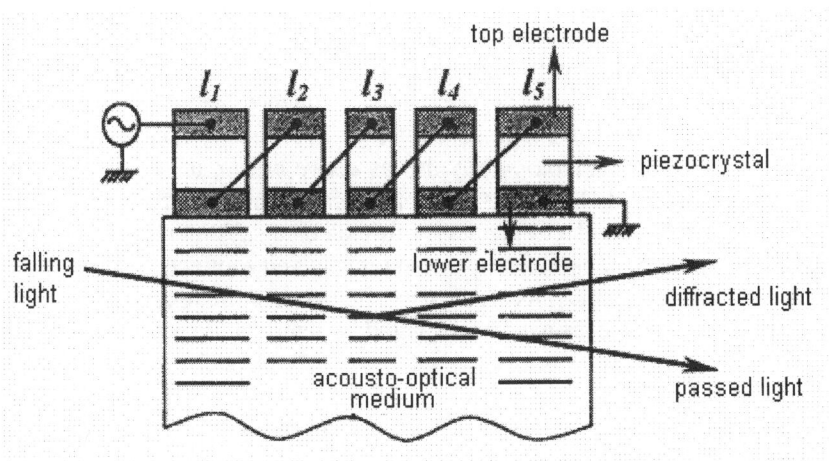

Fig. 2.

A typical piezotransducer consists of a plate glued onto acoustic-optical material. This plate comprises three layers: the top and bottom electrodes and a piezocrystal between them. The glued plate is dissected into several unequal parts, electrically isolated from each other. These sections are se-ries-connected in an electric circuit: the top electrode of one section is connected with the bottom electrode of the next section, etc. Each section is an electric capacity, proportional to its length. The amplitude of the acoustic field is proportional to voltage.

It is impossible to control a technological process, a household appli-ance or a vehicle (a car, a fridge, a kitchen machine) without reliable and cheap *sensors with a microprocessor*. To transform electric signals from sensors for computer processing, there are various interface systems and devices imitating human sensory organs. Computer interface systems are necessary because sensors give an analog electric signal, which has to be digitized. The sensors are not compatible with the computer electrically and constructively (in input velocity, frequency range and signal ampli-tude).

To eliminate this, *ferroelectric semiconductors with high dielectric permeability* are used, in particular in piezotransformers with non-hysteretic control, amplitude modulators, phase revolvers, and delay lines [1, 87].

Polar DE allow one to reduce the size of microwave circuits, because the length of the electromagnetic wave in a dielectric is in inverse proportion to its elastic modulus, and the size of the circuit is reduced correspondingly (electric current is replaced by the piezoeffect) [1, 3, 5, 6, 8, 9].

Ferroelectrics (FE) with high dielectric permeability are used in microwave technology as dielectric resonators, frequency stabilizers, microcircuits, filters, condensers. The substrates of hybrid MW microcircuits act not only as *electroinsulators*, but also as *active media*, in which MW energy is transformed. Thus, the electrophysical properties of FE determine the parameters of an electronic device formed on a substrate [9, 21].

Electrostrictive connection (quadratic inverse piezoeffect) is of practical importance in ceramic DE. It is used in external field-controlled nonhysteretic acoustoelectronic devices with low temperature sensitivity.

Superconductivity was discovered by Dutch physicist Heike Komerlingh at the University of Leiden in 1911. After success in the liquification of helium (He), Kamerlengh Onnes observed that the electrical resistance of a mercury filament drops abruptly to 4.2 K.

Superconductors exhibit zero electrical resistivity and become diamagnetics when they are cooled to a sufficiently low temperature [Burns G. High-Temperature Superconductivity. Boston, Ma.: Academic Press. 1992].

The Josephson effect, discovered by the Nobel Prize winner, is one of the most amazing phenomena in solid-state physics (1965). High-temperature superconductivity [Bednorz J.G., Müller K.A. Z.Phys.1986. Bd.64, 189] enable the use of the Josephson junction in superconductive quantum interferometers, or SQUIDs (Superconducting Quantum Interference Devices). They are used to study the internal thin structure and physico-chemical properties of materials, as well as biological objects.

The sensitivity of SQUIDs is several orders higher than that of the most sophisticated rubidium magnetometers.

"Superconductors have two main applications – magnets and electronics.

"Superconducting magnets can be used in magnetic resonance imaging and separation, magnetic levitation trains. For power utility they are promising for electrical power transmission, motors and generators.

In electronics superconductors are applicable for logic devices and memory cells, field-effect transistors (FET) and Josephson junction (JJ) integrated circuits.

SQUIDs are the most sensitive detectors of magnetic fields. They measure very faint signals of human brain (in the order of 10^{-15} Tl) and heart.

HTSC (high-temperature superconductivity) permits the fabrication of superconducting electromagnetic radiation detectors for over the spectrum from X-ray to the far infrared (IR)" in contemporary devices, which operate at ~150 K.

Josephson junctions (JJ) are used in logical or/and circuits with two and four outlets. A SQUID consists of JJ connected parallel with superconducting inductivities (controlled by a magnetic field) and resistance, which is necessary to create attenuation in resonance contours formed by inductivities and capacities of the SQUID [Zappe H.H., Landman B.S. Analysis of resonance phenomena in Josephson quantum interference devices. *J. Appl. Phys. Jan.* 1978. Vol. 49, p. 344-350].

The optimal choice of the excitation frequency and methods for the self-synchronization of oscillations are considered in § 3.4 and § 3.5 (one can also use varicaps, or nonlinear capacities [1, 9, 11, 12]).

1. Polar dielectrics and piezosemiconductors. Properties and use

1.1. Piezo- and pyroelectrics. Electrets

If an electrically neutral conductor (metal) is placed in an electric field, its surface will be covered by induced free charges, which fully compensate the external electric field. The surface of a *polar dielectric* (in an external field) is covered by induced *connected charges*, which, however, *do not fully* compensate the external field. These "insulators", which may *hold an electric field*, are called dielectrics (DE). The insulating properties of DE are caused by a great quantity $(10^{18}\,\mathrm{cm}^{-3})$ of charge carrier traps. An external constant electric field *polarizes* DE. In metals polarization does not occur due to high concentration of free electrons forming electron "gas" (around positive ions), which screens the external electric field. In DE and piezosemiconductors at high temperature and dynamic excitation, *polarization and electroconductivity in an external electric field* appear simultaneously, being practically inseparable. Under harmonic oscillations in polar dielectrics there is *a "biasing" current* (or capacity current), which characterizes *temporal changes of polarization*. As a result, the surface density of charges (without exiting to the surface) creates an apparent conductivity effect, which is added to the current of free charge carriers [1-3, 5, 9, 12, 18, 23, 24, 29, 37, 46, 61]

The term "dielectric" was introduced by M.Faraday (1839). It originates from the Greek word *dia* ("through") and *electric* and denotes media which *can be permeated by an electric field*, as distinct from metals (which screen electrostatic fields). Comparing the *optical properties* of DE and metals, one should note that the presence of free electrons in metals causes electromagnetic waves to reflect off the surface of metals (which accounts for their typical metallic shine). On the contrary, electromagnetic waves *easily enter DE*. PLZT-type ceramics with an admixture of lanthanum is transparent and is used in protective goggles and windscreens, which automatically change their transparency depending on light intensity.

Dielectrics (DE) are materials (solid and fluid) with low electric conductivity. However, in weak fields all DE have a slight degree of conductivity due to charged impurities, defects and *injection of electrons from the cathode*. The conductivity of DE under a constant electric field is $10^{-10} - 10^{-15}$ $Ohm^{-1}cm^{-1}$. If the strength of the field is increased to about 100 kV/cm, the conductivity of DE increases and they are broken down. In this case, as experimental study shows, *free charge carriers move chiefly along dislocations.*

Electric conductivity of polar DE in contact with metal electrodes may be caused by *injection of charge carriers from the electrodes* (electrons from the cathode or "holes" from the anode). Injection currents in thin-layer MDM structures with a 50-100 μm thick ceramic layer can occur under 15-20 V. *Injection currents in dielectrics* (under conditions of thermal and autoelectronic emission from metal electrodes) are studied by *dielectric electronics.*

Dielectric properties are exhibited not only by crystals, but also by amorphous substances, e.g. plastics (lavsan, polyethylene, polypropylene), glass, organic compounds like beeswax, amber, etc. (under a strong electric field they have piezoelectric properties).

The properties of *crystal* DE are considerably different from the properties of *amorphous* substances. *Amorphous bodies* are *isotropic*, i.e. their characteristics (e.g. electroconductivity) are *the same in every direction.* DE *crystals* exhibit *anisotropic* properties. For example, quartz and sugar crystals have different conductivity, optical and heat properties in different directions, rock-salt crystals (NaCL - sodium chloride) have different elastic properties. The anisotropy of properties is explained by the *polarization* of DE and *translation symmetry* of their crystal lattice.

The main type of physical *"anisotropy"* is *unipolarity*, i.e. the *change of conductivity* with *the direction of the external electric field* (when the poles on its electrodes are reversed). Piezoelectric properties of polar DE are connected with the appearance of opposite charges on the surface of a compressed crystal.

Polarization is the formation of electric dipoles. The simplest dipole is two equal but opposite electric charges q at *l* distance from each other. The electric moment of the dipole $\overline{p}$ equals $\overline{p} = \overline{q}l$ ($\overline{q}$ is a single charge, *l* is the distance between the charges). Vector $\overline{p}$ is directed from the negative charge to the positive charge, i.e. *opposite to the vector of the electric field strength, E.*

Free charges in the conductor and semiconductor move under the influence of an electrostatic field and within 10^{-15} s fully compensate the external field.

Polarization can be "elastic" and residual (irreversible). During the industrial manufacture piezoceramic elements are preliminarily polarized at 15-20 kV.

Some natural crystals: tourmaline, Rochelle salt (found by pharmacist Signette on the inner surface of wine barrels), are characterized by spontaneous polarization.

Polarization occurs when electric charges in a DE are displaced at one tenth of lattice distance (3-5Å). These processes also include deformations of electron shells, ion shifts, changes in the orientation of electric dipoles and "domains". The term "polarization" can also be used for *charge accumulation and separation* on micro non-homogeneities of the structure, *interfaces of materials and phases* in polycrystals.

Polarization also creates *bound charges* on the surface of a DE, which form an inverse electric field (the field of bound charges is smaller than the external electric field).

There are several types of polarization:

electron polarization,

ion polarization (displacement of ions in the crystal lattice in regard to each other),

orientation polarization, i.e. rotation of dipole groups (domains) in the external electric field,

dynamic relaxation polarization (movement of ions as a result of uneven charge concentration along thickness),

interlaminar Maxwell-Wagner polarization (which occurs *at the interface of different materials*).

All types of polarization are found in piezoceramics.

Electron polarization is an elastic (reversible) displacement of electron orbits in regard to nuclei in atoms, molecules and ions. This polarization occurs in all DE in $10^{-14} - 10^{-15}$ sec and can be compared to the period of light oscillations. *Non-polarized ceramics* has *only electron* polarization (because the centers of equivalent positive and negative ion charges coincide).

Ion polarization is a displacement of charged ions in the knots of ion crystal lattice in $10^{-12} - 10^{-13}$ sec. Ion crystals may also have electron polarization, thus such DE have a much larger dielectric permeability (2 or 3 orders) as compared to amorphous polymer materials.

Orientation polarization is most typical for ceramic polar DE with *the ion bond*. Their centers of opposite charges *do not coincide*. Consequently, without an external field the molecules of polar DE have their own electric moment, i.e. are *"rigid" constant dipoles. This asymmetry of crystal lattice explains piezoelectric properties of polar crystals.*

Orientation polarization is a relatively slow (relaxation) type of polarization ($10^{-10} - 10^{-12}$ sec). *Relaxation polarization lags in time compared to the external electric field, which is explained by energy dissipation.* It is energy dissipation that determines the frequency capture and extraction of weak signals from noise in radioelectronics (see § 3.5). The value of relative dielectric permeability ε_{33}^T for industrial ceramic compositions may reach $2 \cdot 10^3$.

Electroconductivity of polar DE

A piezoelectric transducer can simultaneously operate at constant and alternating external voltage

$$U = U_0 + U_1 e^{j\omega t},$$

where U_0, U_1 are the constant values of voltages between electrodes. In this case there is a biasing current

$$I_{cap} = \varepsilon_{33}^T \varepsilon_0 \omega U_1 e^{j\omega t} / h,$$

where ε_{33}^T is relative dielectric permeability, ε_0 is the dielectric permeability of vacuum, h is the thickness, ω is the angular frequency of harmonic excitation (rad/s).

The density of active constant current taking into account only the electron drift is [A/m^2]

$$I_{act} = \varepsilon_{33}^T \varepsilon_0 \tilde{\mu} U_0^2 / h^3,$$

where $\tilde{\mu}$ is the mobility of free charges.

In case of thermoelectron emission from the cathode, the concentration of electrons injected into the DE sharply decreases towards the anode (almost exponentially), creating a diffusion current. The summed current is determined from the formula

$$j(x) = (9/8)(U_0^2 / h^3) + (3U_0 / 8(hx)^{3/2}),$$

where $\tilde{U}$ is dimensionless voltage, x is dimensionless space coordinate, marked off the cathode (normalized by the Debye screening radius).

In this formula the first member coincides with the well-known solution by Mott and Gurney, and the *second member corresponds to the diffusion component of the current.*

Electron mobility for PZT-4 type piezoceramic materials, most widely used in industry, equals $\mu = 0.1 - 0.2$ cm^2/s*V (for aluminium $\tilde{\mu} = 10$ cm^2/s*V, for copper $\tilde{\mu} = 35$ cm^2/s*V).

At room temperature the average speed of electron movement in metals at a given value of constant current is *from 3 to 5 cm/s.* In case of electric breakdown of dielectrics the speed of so-called *"hot"* electrons is 10^7 *cm/s (avalanche breakdown).*

The density of the constant current (to the area of the specimen cross-section [A/cm^2]) is calculated from the formula

$J = nev$,

where n is the concentration of electrons, e is the charge of the electron, v is its speed.

We know the following types of electroconductivity of DE: leakage current, through conductivity current and depolarisation current. If a condenser is connected to a constant voltage electric circuit, the following types of current appear:

- absorption current (caused by the change of the condenser capacity), i.e. an integral characteristics of MDM structure,
- polarization current, i.e. additional directed displacement of unevenly distributed bound charges (charges appearing on the electrodes because of the displacement current create in DE an additional electric field, which is uneven in thickness and partially screens the external field);
- through conductivity current is an integral characteristics of all above mentioned types of current.

Through current is explained by the leakage of charges from electrodes under the influence of environment (or in defective structures of the substrate).

The dependence of electric induction (or polarization P) on field strength E for most piezoceramic materials is nonlinear (hysteresis loop). Dielectric hysteresis in pre-polarized piezoceramics is characterised by a graph (Fig.3), in which P_0 is the residual polarization, E_c is the strength of the inverse coercitive electric field, which eliminates the polarization of the mechanically unloaded piezoelectric. Under mechanic compression (or tension) hysteresis loops change in shape.

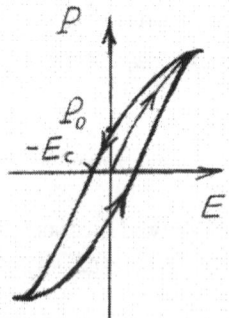

Fig. 3.

In *electrostrictive* ceramics polarization is produced by a strong constant electric field, whereas the additional, essentially lesser, field creates functional displacements. The hysteresis loop is shown in Fig. 4 (deformations are only *positive*). When electrostrictive ceramics is harmonically excited, the piezoelement oscillates at *double frequency* [1, 2].

Electric and mechanical hysteretic phenomena in oscillations and cyclic deformations determine the energy dissipation in piezoelectric transducers and phase shift between the current and voltage (and polarization).

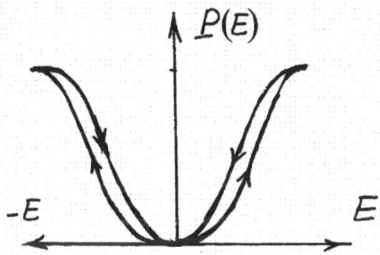

Fig. 4.

Dielectric energy losses are calculated from the hysteresis loop area (Fig.3, 4). The heating of piezoelectric transducers under harmonic excitation is calculated from the function of energy dissipation in the right-hand part of the heat conductivity equation. The integral characteristics of energy dissipation, i.e. tangents of the angles of *dielectric, mechanical and piezoelectric* (coupled electromechanic) losses, are determined experimentally [15].

Electrostrictive ceramics (in the non-polar, paraelectric phase) has a smaller nonlinearity of rheological properties and its piezoelectric modules are almost twice the value of the modules of PZT-4 ceramics. All its functional characteristics are less dependent on temperature. Consequently, piezo- and optoelectronic devices based on electrostrictive materials (including transparent PLZT-type ceramics) are applied in precise optical systems.

Polar DE are divided into two types: (Fig. 5) – *pyroelectrics* (Fig. 5a) and *ferroelectrics (FE)* (Fig. 5b). Ferroelectrics have spontaneously polarized domains with various orientations (Fig. 5b). Around the Curie point (in the phase transformation area) ferroelectrics exhibit anomalies of virtually all physical properties: mechanical, electric, heat and optical.

Nonlinear (step) changes of FE properties are used in various measurement and radioelectronic devices (varistors, varicaps).

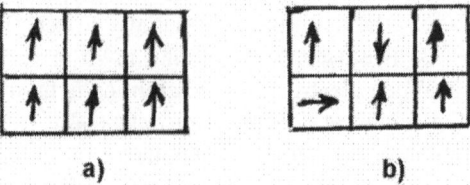

Fig. 5.

External influence	Physical effects		
	Electric	Mechanical	Heat
Electric field	Inverse piezoelectric and electrostrictive, electron injection, polarization, electroconductivity	Piezoeffect, degradation	Heating under constant and alternating fields. Electric aging and electrothermal breakdown
Mechanical stresses	Direct piezoeffect (voltage)	Elastic deformation	Elastic-heat effect (compression leads to heating, tension leads to cooling)
Temperature change	Pyroelectric effect	Thermal expansion	Heat breakdown

Table 2. DE responses to external physical fields

DE properties are explained by their energetic electron spectrum and the width of "prohibited zones" ($\sim 1 - 9$ eV). The lower boundary of "permissible" values of electron energy and the conductivity zone are divided by a prohibited zone (0.9 eV for PZT-4 piezoceramics). Electrons can move from the valence band to the conductivity zone as a result of heating, strong electric fields, light (photovoltaic effect), X-ray and γ-quantum irradiation. Therefore, the division into conductors and DE *is not absolute*, since under certain conditions *a DE may become a semiconductor* (electron injection from electrodes)

DE differ from conductors in a stronger *electron bond in atoms, which does not allow electrons to move freely*. The absence of electroconductivity allows for strong electric fields up to 10^5 V/cm. At high electric field strengths defects emerge and cause the electric breakdown. The electric field strength in metals does not exceed 10^{-3} V/cm.

At inverse *linear* piezoeffect, mechanical deformations are *in direct proportion* to the electric field. Unlike the linear piezoeffect, the *electrostrictive* effect causes deformation proportional to *the square of the external electric field*. Thus the polarity of the electrostrictive deformation does not depend on the vector of the electric field strength. The frequency of crystal oscillations under the *linear inverse piezoeffect* coincides with the frequency of excitation. The oscillations under the *electrostrictive effect* occur with *double frequency*.

In electrostrictive ceramics the coefficient of electromechanical coupling (CEMC) changes with the external electric field from 0.4 to 0.75. This can be used to design piezoelectric transducers controlled by the external field [9, 11, 12].

Material	Relative dielectric permeability ε_{33}^{T}	Piezomodules $(d_{33}/d_{31})*10^{-12}$ [K/N]	Elastic moduli $E_{e\text{-}last}*10^{-5}$ [MPa]	Fatigue limit σ_{-1} [MPa]
Quartz (monocrystal)	4.5	23/7	0.87	-
Barium titanate (ceramics)	15	100/45	1.0	15
Barium titanate with calcium (ceramics)	12	113/50	1.1	15
PZT-4	16	250/100	0.62	30

Table 3. Piezoelectric properties of polar DE [1, 23]

Mechanical losses in UV and radio range (caused by the switching of domains) are reduced. Thus the quality of electrostrictive ceramics (and, correspondingly, the amplitude of displacements at resonance frequency) is approximately 1.5 – 1.8 times as high as that of piezoceramics.

These properties of paraelectric DE, controlled by the electric field, are used in nonlinear MW devices for tuning or increasing the range of functional frequencies of piezoelectric resonators and acoustic-electron generators.

Brittle ceramic materials (as well as cast iron) have poor tensile strength. Compression strength for PZT-4 $\sigma = 200$ MPa, $\sigma_{-1} = 30$ MPa (harmonic oscillations).

Conductors are characterized by the covalent metallic bond. In metal crystals atoms are situated so close to each other that the orbits of outer (valent) electrons overlap and the electrons can move from one atom to another. Electrons in conductors and semiconductors in the conductivity zone are free. The other atoms are viewed as a system of "immobile" positive ions bound by one electron in the valency zone (covalent bond).

Crystal is seen as a unity of ions; each of them loses individual control over *one* or *more* valent electrons. These electrons are common, which leads to a stable ion-covalent bond (e.g. in diamond and gallium arsenide).

A material with a full valency zone and *without free electrons in the conductivity zone* is an *insulator*. Under a strong electric field or high temperature such a crystal may become conductive; as a result, electrons are freed from traps (impurities, structural defects). The transition of electrons from the valency zone to the conductivity zone occurs at very strong electric fields, because for most isolators the width of the prohibited zone is 1-5 eV (for DE local heating in the "point" is 300^0C).

The *electroconductivity of semiconductors* depends on very small changes in the *external electric field and temperature* (or the effect of light), because the width of the prohibited zone is small, about 0.1 eV.

Ion, covalent, ion-covalent and molecular bonds are realised in DE simultaneously. The electric field applied to an ideally pure DE *does not lead to the transport of electrons* (electroconductivity) and causes only *displacement of bound charges (electric polarization)*. Electric, magnetic and mechanic phenomena are largely caused by polarization. Under higher temperature and oscillations in MDM structures and piezosemiconductors, polarization and electroconductivity increase and reveal themselves simultaneously. It is easy to separate them in *quasistatics*: the current exists *synchronously* to the applied external field. DE polarization causes a *biasing current phase-shifted in relation to the electric field*.

MDM and MDS (metal-dielectric-semiconductor) nonlinear structures have a wide industrial application, e.g. in transistors, where FE hysteretic properties are used for controlling n-p transitions (field-controlled dielectric tracts in MW electronics). Even a slight change in pressure, temperature, magnetic or electric field produces a sharp increase in the electroconductivity, especially for thin-film DE. In thin MDM structures charges injected from metal electrodes will drift and diffuse.

Film technology alongside with the direct and inverse piezoeffect uses the electret effect in spontaneously polarized DE. The term "spontaneous" stresses that the crystal is electrically polarized in the absence of an electric field, i.e. represents an electric "battery", a source of electromotive force (EMF).

Apart from constant magnets, there are *constant "electrics"* (*"accumulators" of charges), electrets*, i.e. materials with the properties of a charged condenser. Bound charges under spontaneous polarization create an external electric field. However, unlike constant magnets, bound charges are *compensated* by *free charges in the atmosphere* and in some cases are quickly lost.

Electrets are amorphous or polycrystallic materials which preserve electric polarization caused by an electric field and other manufacturing conditions (electric analogs of the constant magnet). On the opposite surfaces of electrets there are constant non-compensated charges, which exist for hundreds of years. In 1732 S.Grey obtained electrets by cooling melted beeswax, resins and sulphur under a constant electric field (thermoelectrets in modern terminology). In *1839 M.Faraday* wrote about dielectrics, which maintain an electric moment in the absence of an external field. The term "electret" was introduced in 1892 by O.Heavyside. In 1919 M.Eguchi manufactured the first electret transducers from beeswax and a kind of palm resin (microphones, telephones), used in Japan in World War II. In 1953 electrets were synthesized from polymers (Teflon, plexiglass, nylon, polyamide, etc.); in 1956 G.I.Gubkin was the first to manufacture electrets from inorganic materials ($CaTiO_3$, $MgTiO_3$).

V.M.Fridkin and I.S.Zhyoludev studied photoelectrets and their electrophotography [9]. The first electrization of DE with an electron ray dates back to 1973.

In 1970 a electrographic method was developed to charge DE in a corona discharge using liquid contact electrodes. This is possible due to contact potential difference in a strong electric field. Thus, electrets bear different names in different sources: thermoelectrets and electroelectrets.

Since 1962, when the microphone with an electret polymer film was designed [1, 24], the USA, Japan and the USSR have been producing electret microphones and telephones. In 1990 Russia manufactured 2 million items; Japan yearly produces 80 million.

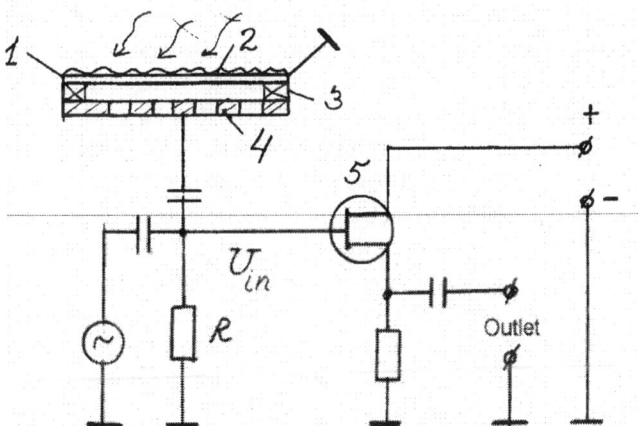

Fig. 6.

Figure 6 presents a layout of the electret condensator microphone: 1 is the metallised electret film (200 μm), 2 is the porous thin felt cover, 3 is the insulation, 4 is the metallic electrode, perforated for the travel of sound waves, 5 is the field transistor. To avoid unwanted signals and leakage currents, the outer part of the microphone, the metallic electrode of the electret film (200 μm) is grounded and contains a layer of felt. The electret film (1) contains an electric charge on the surface facing electrode 4; thus, 1 and 4 form a condenser with an air-gap, which changes as the sound wave travels. The condenser sends an alternating electric signal to field transistor 5, the gate of which also has high ohmic resistance (thus matching the electric impedance).

USW techniques for non-destructive testing (NDT) have been widely used in industry [1, 9, 24]. One of the main parameters tested is the noise level. To measure noise, we use piezoceramic microphones. USW resonance piezoceramic sensors may be used to monitor the frosting of condensers in refrigerators. Piezoelectric devices are also used as timers in refrigerators with automatic de-frosting.

USW welders are used for manufacturing and servicing radioelectronic devices, integrated circuits, microprocessors, and other microelectronic devices.

USW travel according to the laws of geometric optics, applicable for most wave types (including light waves):
- sound reflection and ray refraction on the interface of different media,
- sound diffractions on rough surfaces and at rigid inclusions,
- waveguiding.

The main parameter of USW defectoscopes is the ratio between wavelength λ and the dimensions of the source or obstacle (D). When $D \geq \lambda$, sound travels according to the laws of geometric optics. At 10-30 kHz with $\lambda = 6\text{-}18$ cm acoustic waves behave like optical beams. Modern USW tomography allows one to obtain three-dimensional pictures of solid structures (with layer thickness comparable to wavelength, i.e. several μm).

Electroacoustic, magnetostrictive and piezoelectric transducers (in 100 kHz – 5 MHz range) are used in measurement devices (liquid and granular level sensors, defectoscopes, sensors for measuring the thickness of films and coatings). As compared to magnetostrictive transducers, the oscillations amplitude of piezoelectric transducers is *ten times larger*; their construction is simpler and they *do not require cooling* (at high acoustic powers).

Piezoceramic concentrators have the shape of *concave lenses which radiate converging spherical waves*. At the focus of an acoustic concave lens sound pressure reaches $10^5\text{-}10^6$ Wt/cm^2 at 1 MHz.

USW are used for cleaning materials, machine parts and radio components. Cavitation in liquid media occurs when sound pressure reaches 2-3 Wt/cm^2. When cavities implode, the pressure can be 100 atm.

At 22 – 100 kHz waveguide concentrators (in the shape of metal bars with variable cross-sections) are used to increase the 1 – 1.5 μm movements of the piezoelement up to 15 – 30 μm (USW welding).

1.2. Piezoelectronics in automobiles

Your tyre will tell you about its temperature and pressure!

Lightning reaction to changes of air pressure in tyres is vital, not only for saving fuel. It can save your life, too! Yet we often gauge tyre pressure by eye.

To keep the driver informed if the pressure in tyres remains normal, Michelin designed a system of pressure monitoring. Each tyre contains a piezoelectric pressure sensor and an annular antenna. Another antenna, placed on the axis, sends an electric signal to the rotating sensor and receives the response signal. The response signal is decoded by a microprocessor and compared to the required value. If pressure dangerously falls or the tyre overheats, the dashboard flashes a red light, which indicates the punctured tyre.

The results of experimental research to build telemetric systems are given in [34] (see § 3.5 and Appendix 5).

Piezoelectric motors are used:
- for cooling electronic devices of board diagnostics and engine radiators,
- in motors for rotating wipers (cleaning car windows),
- in devices controlling airflow in conditioners (ventilators, shutters),
- for ejecting and rewinding cassettes in tape-recorders,
- for turning wind mirrors.

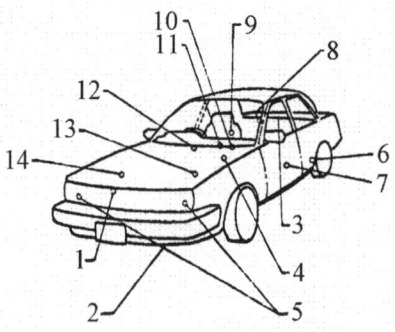

Fig. 7. Piezoelectric motors and sensors in automobiles (1 – radiator cooler, 2 – clearance measuring device, 3 – rear window wipers, 4 – windscreen wipers, 5 – headlight wipers, 6 – fuel pump, 7 – window openers, 8 – telescopic antenna, 9 – automatic seat position regulators, 10 – conditioner, 11 – remote cabin temperature sensor, 12 – actuator (for ejecting cassettes from the tape-recorder), 13 – fuel control)

1.3. Techniques for information measurement

1.3.1. Remote temperature sensing

Pre-polarized piezoceramics (PZT-4 type) is also a good pyroelectric. In it, *heat energy is directly transformed into electric energy. Time temperature gradient* changes the orientation of polar molecules (or polarized domains) or the distance between atoms. This causes a *change in spontaneous polarization*, which leads to the appearance of non-compensated charges on the DE surface. If a polar DE is connected with a measuring device, charge leakage creates a pyroelectric current (or voltage in an open circuit). A 10 K temperature change creates a voltage of about 1 kV on a thin (1 mm) tourmaline plate.

The change in polarization is calculated from the formula $\Delta P = \gamma \, \Delta T$, where ΔT is the change of temperature in time, γ is the pyroelectric coefficient.

A 1 mm tourmaline plate ($\gamma = 1.3 * 10^{-5}$ C/m^2*grad) registers temperature changes of 10^{-5} K. If this plate is heated by 10 K, there appears a charge with the surface density of 0.5 μC/m^2, which corresponds to the potential ≈ 1.2 kV (for barium titanate $\gamma = 0.5*10^{-3}$C/m^2 * grad).

Pyroelectric crystals are in particular used for remote temperature sensing in IR (night vision devices). They work at wavelengths ranging from 3 to 12 μm and in the submillimetric range with 50-200 μm wavelengths. IR radiation in the submillimetric range easily penetrates textile materials (fiber polymer structures) and is used in airports to search for metal articles in passengers' clothes and luggage. Devices of this type use laser lighting [24, 59, 61, 65, 66, 85-88].

Pyroelectrics are very sensitive: in the dark they react to human movement at a distance of 150 m. In the simplest measuring set consisting of standard television equipment, microprocessors and computers, pyroelectric receivers produce a 2-dimensional (20-point) image of temperature fields on the screen. These systems may be used for the sample test of refrigerators to locate cold leakages ("cold bridges") and monitor various technological processes. For example, IR photography of buildings in Sweden and France helped to reduce their heating costs [88].

Remote temperature sensors are used in construction for testing heat insulation of walling and piping, even underground.

Pyroelectric properties are exhibited by the following ferroelectrics (FE): triglycinesulphate, niobates and lithium tanthalate, as well as PZT-4. CdS-type crystals, lithium sulphate, etc. make up a second group; film

PVDF polymers constitute a third. Thin, elastic and strong polymer films seem to be the most promising for application [20].

Pyroelectrics are used for analyzing not only temperature, *electromagnetic* and *ionizing* radiation, but also *shock waves* and *chemical composition*. They receive radiations in a wide range, from centimetric waves to X- and γ –rays, including the visible part of the spectrum. Temperature is usually measured in the 8-12 μm range *"second atmospheric window"*. Mist (moisture), CO_2 and other gases absorb 8-12 μm waves; thus, measurements in mist are made in the submillimetric range (with the wavelengths $\lambda = 70$-200 μm).

Pyroreceivers can consist of many sensitive elements forming a pyroelectric matrix (10^3-10^5 elements).

A pyroelectric receiver has a resistance of about 1 GOhm. To eliminate the influence of heat "noise" and background, the signal from the pyroreceiver is sent to the gate of the field transistor (with a considerable input and neglible output resistance between the "sink" and the "source") (see Fig. 6).

Pyroelectric image transducers transform heat radiation images into electric signals or TV pictures. A heat image in a pyroelectric vidicon (camera) is projected onto the pyroelectric plate (target) and creates an electric pattern (distribution of pyroelectric charges). The electron ray scans the pyroelectric pattern and transforms the electric signal. The contrast of the heat image is fractions of degrees, and the definition quality is several hundreds of lines per target.

Pyrovidicons are used for remote testing of homogeneity and insulation in electric devices and machines and for automated technological testing of electronic components under voltage (resistors, condensers, very-large-scale integration circuits VLSI).

Pyroelecric transducers are used to measure temperature, specific heat, heat conductivity and heat exchange. Their limit sensitivity is 10^{-7} K.

Pyroelectrics can be used by jewellers *to identify precious stones* (diamonds). A pencil-like device with a stabilized heat source and a pyroelectric touches a diamond with its tip and identifies glass by the time of "leakage" of the heat impulse (the heat conductivity of diamond is *thousands of times* higher than glass and twenty times higher than silver). Thus, thermal physical properties are the most informative characteristic of a material.

Industrial pyroelectric heat receivers have a high reaction speed and work at 10 MHz. A pyroreceiver *does not require power supply* to function (which is necessary for a photodiode). There are no leakage currents in multi-element pyroelectric devices, because pyroelectrics are high ohmical DE.

The diameter of the observed "spot" d on the surface is calculated from the formula $d=10^{-2}l$, where l is the distance to the object. When l = 25.0 cm, the "spot" diameter (in which temperature is precisely measured without optics) d = 2.5 mm. Error at the distance up to 10 meters is about 1K.

A standard system for transforming heat 2-dimension images contains a monolithic raster of pyroreceivers and operates in the matrix addressing mode. There is a line network on one surface of the FE plate and an orthogonal system on the other surface. Each element of the matrix is formed at the intersection of the line and column. Pyrosignals are electronically scanned by integrated circuits.

The use of conventional *radioelectronic VHF devices* is *limited in principle*. It is impossible to send a signal *directly* at short distances without electromagnetic noise in the environment. The object can be observed only using *sophisticated radars*. These drawbacks are eliminated with IR pyroelectric devices [85-87].

IR radiation belongs *neither* to electromagnetic *radio frequency* waves, which can be received by antennae, nor to *visible light*. Unlike the other wave types, these vibrate at the eigenfrequency of body tissues and can be received by a living being. IR waves may be compared to *"travelling"* waves on the *surface of water* (their length is measured by the distance between the humps).

IR radiation with amplitude and frequency modulation is used for wireless transmission of speech or music. Thus, we may build *wireless tap-safe telecommucation systems* working in 150 m range.

IR range is conventionally divided into three intervals: *short-distance* (0.7-2 µm), *mid-distance* and *long-distance* (15 µm-1 mm). "Heat" radiation felt by a human hand near a welding tool belongs to the mid-distance interval (6-15 µm). Indium antimonide-based IR receivers can measure radiation temperature at the distance of *many kilometers* accurate to 0.1°C (with optical lenses). Pyroreceivers give a real-time heat "picture" on a computer display, with each point having its own temperature. This "heat" picture of the object may be "seen" *at night in reflected starlight*.

In the *mid-distance* IR interval (6 to 15 µm wavelength) there are relatively cheap sensors, used in industry. They are called *"passive"* optical pyroreceivers, because they only register natural radiation, unlike active pyroelectric receivers with lighting. These receivers are used in alarm systems, automatic lighting, temperature control devices, etc.

Short-distance IR devices (*light-emitting diodes, photodiodes and phototransistors*) can transmit large amounts of information.

Sources of IR radiation include INC (incandescent lamps), FL (fluorescent lamps), LA (semiconductor lasers), LED (IR light-emitting diodes based on GaAs).

The couple silicon *light-emitting diode* – silicon *phototransistor* seems to be the most reliable. Standard detector PID11 (Siemens) works at 4.5 V, consumes 0.4 mA and has an output resistance of 2 kOhm [37, 88].

The maximum value of photodiode relative spectral sensitivity (RSS) corresponds to the width of the prohibited zone in the crystal of the electron-optical converter (EOC).

Semiconductor diodes are based on Si, Ge, GaAsP, GaP (GaAsP is a heterojunction photodiode). The maximum RSS for a Si-photodiode corresponds to the wavelength $\lambda = 0.975$ μm, for GaP $\lambda_{max} = 0.55$ μm [Moss T.S., Burrel G.J., Ellis B. Semiconductor Optoelectronics. London: Butterworths Sci.Publ. 1973].

Semiconductor lasers can transmit information by switching on and out at *1 billion bytes per second*, whereas *avalanche photodiodes* can receiver information at this frequency. This optical fiber communication line (OFCL) consists of 0.13-0.3 mm fiberglass. Large amounts of information are transmitted without amplification at hundreds of kilometers (dozens of TV channels or 15 20 thousands of telephone talks at the same time).

Fiberglass optics is used in laser scalpels (for brain surgery, gastric ulcer and tuberculosis treatment); lasers grow crystals, cut metals and work at wavelengths from UV to IR. In the mid-1990 s, Russia was the first to create the "quantoscope" based on standard electron-beam tube technology in which, however, the electron ray creates coherent laser radiation in a semiconductor plate. "Blue" light identifies the cancer cell, and "red" light boosts oxygen metabolism and the cell destroys itself. This method helped cure more than 10 thousands of cancer patients around the world.

OFCL has the following features:
- the absence of electromagnetic attenuation, "inducing" and leakage currents,
- superior chemical, mechanical and radiation stability,
- combination of various physical effects: *electrostrictive, acousto-electric, elastic-electric* (at 40 MPa the width of the prohibited zone changes).

Due to this, optoelectronic systems based on transparent monocrystals ($BaTiO_3$, lithium niobate) and PZLT ceramics process information in the simplest way without losses.

Determining chemical composition and purity control are one of the practical aims in industry. Spectral luminescence is a much simpler alternative to conventional methods of analysis. It helps to determine chemical composition, crystal structure, phonon spectrums, the type of ions in admixtures and their valency [Mesyats G.A. et al. On the possibility of refining energy levels in solids. *Dokl. Akadem. Sci.* 1994. Vol. 339. No. 6, p.757-760 (in Russian); Mikhailov S.G., Solomonov V.I. Impulse cathode luminescence in diamonds. *Optics and spectr.* 1996. Vol. 80. No.5, p. 781-784 (in Russian)].

1.3.2. USW defectoscope

This device is meant for non-destructive detection of internal and surface defects in the elements of machinery, radio- and microelectronic products. Its general principle is based on USW reflection and dissipation at cracks, cavities and in heterogeneous structures. The defectoscope radiates waves steadily or in impulses at 0.5 – 25 MHz.

Elastic waves are introduced into the tested sample by:
- dry contact technique,
- liquid "lubricant" contact technique,
- *non-contact technique* through an air-gap with an electromagnetic detector (an alternating magnetic field interacts with *eddy induction Foucault currents* induced in the surface layer of metallic parts).

1.3.3. Non-destructive testing

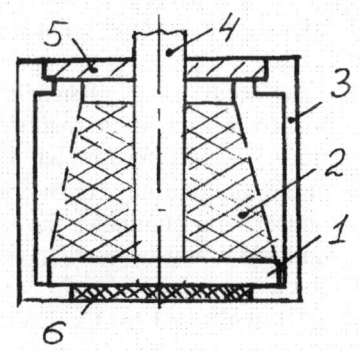

Fig. 8.

A roughly sketched cross-section of a defectoscope is shown in Fig. 8 (1 is the piezoelement, 2 is the damper, 3 is the casing, 4 is the wire of the electric information processing circuit, 5 is the insulating insert, 6 is the protector with the radiating surface).

The echo technique

After sending about 10 periodic acoustic impulses into the sample, the device registers the intensity and receipt time of the reflected echo signals.

1.3.4. "Shadow" technique

The operation principle is demonstrated in Fig. 9, where 1 is the generator, 2 is the wave source, 3 is the sample, 4 is the USW receiver, 5 is the electric signal transducer, 6 is the indicator.

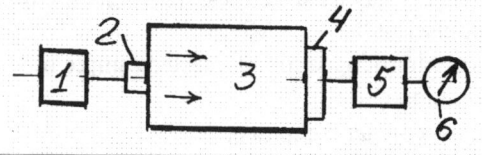

Fig.9.

1.3.5. Resonance technique

This technique is used for measuring the thickness of separate layers in composite constructions from metals, glass, ceramics and other materials of high mechanical quality and for locating failures and lamination zones in bimetals (one-sided access).

A piezoelectric transducer sends waves of steadily changing frequency into the sample. At resonance, with a whole number of half-waves along the thickness, space stationary waves are formed. This reduces the input mechanical impedance of the sample (the acoustic resistance of the external load on the PZ transducer) and eliminates the reactive component of wave resistance. The amplitude of elastic oscillations in the piezoelement sharply increases and the current in the electric circuit of the oscillation generator changes, which is registered by the indicator. Thus, the layer thickness is measured by a whole number of half-waves.

1.3.6. Impedance technique

The principle of the USW defectoscope is shown in Fig. 10 (1 is the metallic deflecting attachment, 2 is the radiating piezoelement, 3 is the sound-conducting plexiglas bar, 4 is the dynamometric piezoelement, 5 is the metallic layer, 6 is the glue layer, 7 is the substrate glued to metallic layer 5, 8 is the light-alloy contact tip with a wear-proof "mushroom" in the centre.

If the casing (metallic layer 5 or other material) is rigidly joined to the substrate, the mechanical system (the composite) performs elastic oscillations as one whole, and the input mechanical impedance is maximal. In this case the reaction affecting the piezoelement is the strongest. In the absence of glue the rigidity of the thin plate is considerably lower and the reaction is 2-3 orders of magnitude weaker, as registered by the electric circuit.

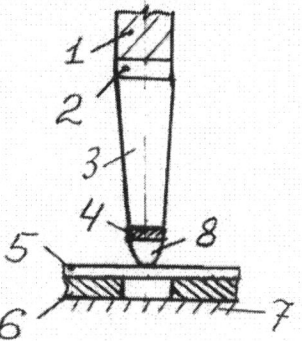

Fig. 10.

To use USW techniques for the non-destructive testing of various-shape parts, one can turn to the algorithm of computer material identification [Gunarathne G., Christidis K. Material characterization *in situ* using ultrasound measurements. *IEEE Trans. on Instrumentation and Measurements.* 2002. V. 51, No. 2, p. 368-373].

1.3.7. Electret effect

Thermoelectrects are fabricated by heating DE in a strong electric field. A charge is also formed when an electret is made by thermostimulated injection of cathode electrons or anode "holes".

Polarization ("charging") of electret polymer films is achieved by exposure to high-energy electrons and to gas discharge. One can use other sources of external energy: radioactive or X-ray radiation, light energy (photoelectrets [9]). The friction of certain pairs of materials (donors and acceptors) causes a polarization of polymer films and tiny (50-300 μm in diameter) particles (triboelectrofication). Triboelectric "charging" may be applied to powder colouration of various materials, such as metals, textiles and wood (by the electrostatic precipitating of charged particles on a grounded metal plate with subsequent thermal treatment). This technology can be used for a uniform thin coating of slots and for inter-groove insulation in electric motors.

Thermoelectrets are mixes of fusible amorphous materials (waxes, bitumens, resins), glass, various monocrystals: oxides, fluorides, etc. – PZT-4 ceramics (unlike PZT-4-based piezoelements, electrets have no electrodes) – polymers – PVDF (polyvinylidene fluoride).

Photoelectrets are DE with high photosensitivity and *low dark conductivity*. The electric charge in photoelectrets depends on the thickness distribution of the electric field, as well as on light distribution. Charge carriers in lighted areas are released due to the photoeffect and drift into the DE, finding their places according to the lighted and "dark" zones (settling on "traps" and forming a homocharge). As a result, after the electric field and light source are shut off, there is an image near the surface, which can be read by the electron ray or "developed" by colouring powder. The particles of the powder will be deposited on the charged areas of the photoelectret due to electrostatic forces. The photoelectric image can be "wiped off" by a strong electric field or continuous light exposure [85, 88].

Photoelectrets can be monocrystal (selenium, zinc oxide, zinc and cadmium sulfides and selenides) and polycrystal (sulfur, anthracene and naphthalene). They are often used in electrography ("dry" electrophotography).

Electrography differs from photoelectret polarization described above. An electrophotosensitive conducting plate, coated with a thin layer of a photoelectret, is pre-charged in the dark by a corona charge. The image to be reproduced is projected on the plate. The charges in lighted areas relax and produce a hidden electric picture, which is transferred onto paper with pigment powders.

Microphones are the most wide-spread industrial application of electrets [24, 74]. Electret microphones have a sensibility of 0.1 mV/μbar and work at 500 – 3000 Hz.

Electrets are used in the following electromechanical transducers:
1. Electrophone pickups, in which the needle is connected with an electret microphone (or based on the piezoeffect in electret films).
2. Hydrophones, vibroacoustic sensors of pressure oscillations (work at 1 Hz-15 kHz).
3. *Sensor switches* which react to the smallest touch of finger, as *even a slight shift of the electret diaphragm induces electric fields of 1-100 V.*

Electrets are used in up-to-date technology for recording signals and images. Electrostatic recording of information (similar to magnetic tape recording) is made on polymer films. Barium titanate or lead zirconate titanate-based ceramics can also be used in electromechanical transducers and electric memory devices. The phenomenon of photoconductivity in ferroelectrics and semiconductors is used in optoelectronics to record holograms.

1.3.8. Posistor effect

Thermistors with a high positive temperature coefficient of resistance (PTC), i.e. posistors, were discovered in 1955, as a result of research into the electroconductivity of semiconductor barium titanate-based *ceramic* [9, 11, 12].

Unlike *thermistors* (resistors with a negative temperature resistance coefficient (NTC) of 3-5%/K, posistors have a PTC of 100 % /K upwards.

The temperature when resistance is increased (10^5 times) coincides with phase transformation temperature (the Curie point) and may be easily varied over a wide range. Upon introducing a *microscopic* amount of admixtures into "pure" barium titanate ceramics its electric resistance *drops by 10 orders*, and it becomes a semiconductor. Normally monocrystal semiconductors become less resistant when heated, whereas the resistance of semiconductor ceramic barium titanate at *80-120 ^0C* (phase transformation) increases $10^4 - 10^5$ times. Thus, the material came to be used in devices which are sources and regulators of temperature at the same time (thermal relays).

An anomalously high PTC in $BaTiO_3$ ceramics is put to the influence of inter-grain layers, which are highly resistant. That is why there is no posistor effect in barium titanate monocrystals. Interface layers at the boundary between crystals are sensitive to mechanical pressure. When $BaTiO_3$ is heated to the Curie point, its crystal lattice undergoes significant changes. These cause considerable mechanical forces, which may affect the inter-crystal layer and bring about a PTC-transition.

Posistors are used in current terminators and stabilizers, for automatic regulation of energy transformation in current and power amplifiers, in electric circuits for temperature compensation. The anomalous properties of semiconductor barium titanate can be used for automatic thermal stabilization. A ferroelectric in an electric field maintains a given temperature regardless of the temperature of the environment (a thermostat). Automatic temperature stabilization is achieved with a nonlinear ferroelectric element, a TANDEL, in which dielectric losses decrease as temperature rises (at a certain phase transformation temperature). Under alternating current it preserves the given constant temperature. Automatic temperature stabilization can be also achieved with ferroelectric elements in resonance systems (MW resonators).

In home appliances posistors were first used for temperature compensation, in thermal relays and in non-contact switches (without moving parts typical of mechanical contacts), as well as in safety-fuses, level sensors and start-up circuits for electric motors.

When a barium titanate-based semiconductor is subjected to mechanical pressure, it demonstrates a *piezoresistive effect* (i.e. may be used for *tensoresistance*). Therefore, posistors can be introduced into supersensitive noise meters and sensors of pressure, vibration and other mechanical characteristics.

1.3.9. Physics of piezoelectric transducers

Effect, properties	Functions, piezoelectronic elements
1	2
Nonlinearity of DE parameters	Multiplication and transformation of frequencies. Transformation of the shape of the electric signal. Mixers. Broadband and selective amplification. Transformation of constant current signals into an alternating current. Electric field sensors. Electric signal modulation (amplitude, phase, frequency). Amplitude detectors. Parametric and compensation stabilizers of current and voltage. Controlled RC-structures. Tunable correcting automatic circuits. Oscillation autogenerators. Non-contact relays and triggers.
Hysteresis	Bistable and multistable memory devices. Logical elements. Static registers. Distributors. Fast amplifiers. Impulse formers.
High dielectric permeability Piezoeffect	Small-size capacities. Electric field concentrators. Piezotransformers. Electrostatic relays. Phase revolvers. SAW devices. Resonators. Frequency-selective filters. Current and voltage transformers. Sensors for transforming mechanical energy into electric energy and vice versa. "Solid-state" indicators and screens (quantoscopes). Actuators. Adaptive optical devices. Electromechanical and electrooptical devices for processing optical information.
Inverse electrocontrolled piezoeffect	Controlled (panoramic) frequency-selective circuits. Controlled delay lines. Amplifiers, modulators, memory devices. Logical elements (transducers of analog signals into digital signals). Actuators.
Direct tensocontrolled piezoeffect	Sensors and electromechanical transducers.

Effect, properties	Functions, piezoelectronic elements
1	2
Pyroelectric effect	Transducers of heat energy into electric energy. Solar batteries. Heat sensors, bolometers. Night vision devices.
Thermodielectric and thermoresistive effects	Temperature stabilizers. Temperature sensors. TANDELs, thermostats. Posistors.
Piezoresistive and piezodielectric effects	Amplifiers. Electromechanical and physical sensors.
Electro- and piezooptical effects	Modulator and scanning devices in optical quantum generators, Measuring, logical and memory optronic devices.

Table 4. Functionality of ferroelectrics (FE) [1, 2, 9]

Characteristic	Construction			
	Semiconductor-based micromodules	Film semiconductor circuits	Monolithic semiconductor circuits	Monolithic dielectric circuits
Packaging density of elements, cm^3	2-3	90	$3.5 \cdot 10^3$	$5 \cdot 10^3 - 10^4$
Average work time, hours	$10^4 - 10^5$	10^5	10^6	$10^6 - 10^7$
Minimal power consumption (to 10^7 Hz), W	$3 \cdot 10^{-3}$	–	10^{-3}	10^{-5} (performance index = 0.6 – 0.9)
Minimal supply voltage, V	5-10	–	1-3	10^{-4}
Working temperature range, °C	from -65 to +125	from -55 to +125	–	from -200 to +750
Radiation resistance	–	–	10^{14} neutron/cm^2 (MIS-structures)	10^{18} neutron/cm^2 (ceramics)

Table 5. FE and semiconductor application [1, 9]

2. Multifunctional piezoelectronic devices

2.1. Piezoceramic transformers (PT)

The combined use of direct and inverse piezoeffects allows one to separate input and output electric circuits. The frequency characteristic of a PT has limited selectivity, and maximum performance can be obtained only *in the resonance mode*. Thus, PT are sometimes used as *filters,* whose conversion efficiency may reach several hundred [1].

A PT is a monolithic (or two glued together) piezoceramic pre-polarized plate with input (*excitation*) and output (*generator*) sections. There are also current transformers.

The transformation coefficient of PT is more than 1000, which gives up to 10 kV of output voltage. The specific capacity of PT is 1-10 W/hour with 90-98% efficiency.

2.2. Field-controlled piezotransducers (FCP)

Resistances, inductivities and capacities may complement a non-controlled *dielectric resonator*, which allows us to change amplitude, phase and frequency characteristics. A nonlinear ferroelectric capacity (varicap) gives the most successful combination.

The response of a piezoelement to external electric influence is determined by the following factors:

1. *The lag* of the input circuit nonlinearly depends on *capacity*. When the output resistance reaches a finite value, there is a transition period in the feed circuit. The input capacity of a piezoelement may be a few picofarads to several nanofarads.
2. The DE electric field time. Its working response is determined by the specific time constant $\tau = \rho_{el} \cdot \varepsilon_{33}^T \varepsilon_0$, where ρ_{el} is the specific electric resistance. In ceramics ρ_{el} varies from 10^{10} to 10^{16} Ohm·cm, whereas $\varepsilon_{33}^T = 10^2 \text{-} 10^4$; thus, τ may range from 10 to 10^4 s ($\varepsilon_0 = 8.85 \cdot 10^{-12}$ [F/m]).

Engineers use the formula
$$\tilde{\tau} = RC,$$
where R is the electric resistance, C is the piezoelement capacity.
3. The time constant characterizing the inertia of the LRC oscillation circuit is
$$\tau_1 = Q / \pi f_p.$$
In high-quality piezoceramics $Q = 10^2\text{-}10^3$, and constant τ_1 equals 10^{-4}-$5 \cdot 10^{-2}$ s.

A universal FCP can be used for *amplifying, transforming and modulating electric and mechanical signals*, as well as for performing *logical functions* and as a *memory element* (Fig. 11). This device consists of equally polarized resonators 1 and 2, connected by element 3. Resonators 1, 2 and element 3 are fashioned with electrodes 4, 5 и 6. This construction may function as a controlled PT (Fig. 11).

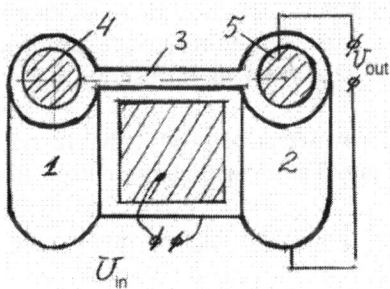

Fig. 11.

FCP are summing devices, parametric current and voltage stabilizers, piezosemiconductor voltage amplifiers, piezoelectric frequency multipliers, multiple reference frequency generators, frequency-selective elements, amplitude and phase modulators.

The dependence of dielectric and electromechanical constants of piezoelectric materials on the controlling electric field allows them to be used in various energy transducers. Specifically, the dependence of resonance frequency and the distance between resonance peaks on the electric field is used in frequency-selective systems with tunable frequency and pass band width.

Current control is used in amplitude modulators, modulator amplifiers and transducers. The introduction of feedback helps to make stabilizing devices, autogenerators, and is often used in various electronic circuits.

2.3. Computing devices

Ferroelectric recording devices (RD) have a higher output signal than ferromagnetic RD. Information can be read without destroying the memory device. It is preserved in FE even when the power supply is switched off. The use of a high-frequency carrier field in FE (in the resonance mode) helps to separate write and read channels, minimize losses and increase the output signal [1].

In the case of an amplitude output, the FE is depolarized when recording "0" and polarized when recording "1". In the case of a phase output "1" is represented by the "zero" phase of the output signal, "0", by the 180-degree phase. The superposition principle allows one to produce not only 2- or 3-stable, but also 6-stable RD.

A RD unit-cell is built as a typical resonator. Information is recorded through input 2 by sending impulses of corresponding polarity, with an amplitude exceeding coercitive voltage (at coercitive voltage induction is zero). To read information, a high-resonance-frequency signal is sent to input 1. The signal is read from the measuring (unloaded) resistor R_u at output 3.

The advantage of a two-electrode RD is that its write and read circuits are electrically separate. The matrix placement of electrodes is the most convenient.

Silicon technology gives the highest integration and reliability of microelectronic devices. Metal-silicon nitride-silicon oxide-semiconductor structures (MNOS) are the basis of VLSI and electrically reprogrammable memory devices with high capacity.

Industrial MNOS consist of dielectric SiO_2 and silicon nitride (SN) S_3N_4 as the memorizing medium. Recording in such memory elements happens due to the entrainment of free charge carriers by SN "traps". This charge is maintained for a year. When recording electric voltage is applied, electrons are injected through SiO_2 into SN film and entrained by the "traps".

The limited number of switching cycles ($n \leq 10^5$) is due to degradation in MNOS, which lead to breakdown. Degradation processes are determined by the fact that "traps" are distributed very unevenly along the DE and SN thickness: trap concentration near the metal electrode N_{DE} is $\sim 2.5 \cdot 10^{19}$ cm^{-3}, and in the SN film $N_{SN} \sim (6 \div 9) \cdot 10^{19}$ cm^{-3} (1.8-3.5 times as great). Due to this difference, when MNOS layers are *relatively* thin, the electric field becomes very strong (hundreds of times stronger than the original *external* field) and therefore the material degrades.

Acoustoelectronics (AE) is one of the newer fields in electronic engineering [21]. AE devices are used in computing, automatic control systems and measurement technologies. They are used in information processing devices for transforming the time, amplitude, frequency and phase of acoustic and electric signals, as well as for integrating, coding, convolution, and correlation (acoustoelectronic generators and amplifiers with a gain factor of 75 dB/cm at frequency f = 30 MHz and 75 dB/cm at f = 60 MHz, filters, signal processing devices (convolvers), phase revolvers). In comparison with their radio counterparts, AE devices are considerably smaller and thus can be well combined with integral circuits. AE dielectrics help to separate the input and output of piezoelectric transformers, as well as work under heat and radiation. They also offer the following advantages: they have a very thin spectral line, tune and synchronize frequency with an external electric field, and produce various signals, from harmonic to impulse, lasting a few nanoseconds.

3. Calculation and optimization of PZ transducers

3.1. Permissible values of current and electric field strength in polar DE

The maximum permissible current value depends on maximum strain without destruction [2]. The permissible strain *depends on the mode of oscillations*. In a quartz crystal at ~ 100 MPa tension, at permissible stress and longitudinal oscillations, the maximum relative deformation of the resonator is $[\varepsilon] \approx 0.0013$. The permissible current density is $I = 10$ mA/cm^2. Thickness oscillations may permit the current density $I_p \approx 120$ mA/cm^2 [2].

At the inverse piezoeffect (under an external electric field) a piezoelement develops internal mechanical stresses, which can be calculated from the formula

$$\sigma_{mech} = \frac{\varepsilon_{33}^T \varepsilon_0 E_z^2}{2} \ \text{[Pa]},$$

where $\varepsilon_0 = 8.85 \cdot 10^{-12}$ [F/m] is the dielectric constant, ε_{33}^T is the relative dielectric permeability, $E_z = -\dfrac{2V_0}{h}$ is the electric field strength ($V = \pm V_0$ are the electric potentials on the electrodes of the piezoelement), h is the DE thickness.

In the one-dimensional case, taking into account pondermotive forces, the mechanical stresses σ_z are calculated from the formula

$$\sigma_z = \sigma_z^0 + \frac{\varepsilon_0(\varepsilon_{33}^T - a)}{2} E_z^2 + d_{33} E_z E_{elast} \qquad (a)$$

where σ_z^0 are the mechanical stresses in the absence of an electric field, a is the constant characterizing the change in dielectric permeability as a result of deformations (for polymethylmethaacrylate $a \cong 4$), d_{33} is the piezomodule, E_{elast} is the Young's modulus.

In expression (2) one should take into account the concentration of mechanical strains, the electric field strength and temperature near the micropeaks on the metal electrodes (see 3.8). The experimental data show that local temperature (point) near the micropeak may be 300^0C, whereas the integral temperature of the dielectric diode is $\sim 30^0C$.

When MDM structures are instantly destroyed, virtually all energy (98% experimentally ascertained) is transformed into heat and changes the polarization of the dielectric layer. Due to the high speed of heating the material has no time to expand, which brings about very high mechanical stresses, spreading as *shock waves*. (Impulse stresses of $\sim 10^{-7}$ s may reach several GPa.) The interference of shock waves leads to the localization of *tensile and shear* strains (causing local destruction).

The results of the radio-acoustic studies of ductile metals (Cu, Al, Fe, Ti) under 2 mega-erg/V electronic radiation for 30 ns have demonstrated that destruction occurs at the planes where falling and reflected stress waves are super-imposed along *grain boundaries* [Steverding B. Fracture by super-imposing stress waves. *J. Appl. Phys.* 1972. Vol. 43. N 7. P. 3217-3219]. *The destruction of polar DE* is also determined by the electron subsystem [Makeyev S.I. et al. Thresholds and some regularities of destruction in piezoceramic materials. *Sov. Phys. – Techn. Phys.* 1983. Vol.53. No.7, p.1381-1383; Wiessburd D.I., Balychev I.I. Destruction of solids due to superdense excitation of their electron subsystem. *Letters to J. Exper. Theor. Physics.* 1972. Vol. 15. No. 9, p. 537-540].

At uneven local heating near surface structural imperfections, there are thermoelastic tensions in metal-dielectric contacts. Taking into account temperature expansion, the dynamic gains of dynamic mechanical stresses are determined by the formula

$$\Delta\sigma_{mech} = -\alpha K(T - T_0) - \frac{a\varepsilon_0}{2}E_z^2 + \rho v_e^2 \frac{\partial U}{\partial x}, \quad (b)$$

where α is the temperature expansion coefficient, T_0 is the initial temperature, ρ is the density, v_e is the velocity of tensile deformation waves (along the bar axis), $dU/dx = \varepsilon_x$ is the relative longitudinal deformation (of the bar), K is the temperature constant.

The last term in the right-hand part of formula (b) is obtained from the Hook's law $\sigma = \varepsilon E_{elast}$. Formula (b) does not include the mechanical stresses $\Delta\sigma_{mech}^p = d_{33}E_z E_{elast}$. Taking into account the well-known expression, the velocity of elastic waves is

$$V = \sqrt{\frac{E_{elast}}{\rho}} . \quad (c)$$

When the cross size of a crystal DE greatly exceeds the transmitted wave length in plastic plates, the velocity of thickness acoustic waves is

$$V_c = \left(\frac{E_{elast}}{\rho} \frac{1-\mu}{(1+\mu)(1-2\mu)} \right)^{1/2} , \quad (d)$$

where μ is the Poisson's ratio.

The velocity of longitudinal waves in thin plates is determined by the formula

$$V_l = \left[\frac{E}{\rho(1-\mu^2)} \right]^{1/2} .$$

Dynamic mechanical stresses in polymers and glasses were calculated by formulas (a), (b), (c) and correlated with experiments [Gering G.I., Yeliseyev N.A. Acoustic phenomena in high ohmic dielectrics subjected to high-current electron beam radiation. *Sov. Acoust. J.* 1989. Vol. 35. No. 2, p. 240-244].

The permissible value of mechanical stresses $[\sigma_{mech}]$ for PZT-19-based piezotransducers = 30 MPa at the field strength $E_z = 10^5$ V/m.

The ponderomotive force $\overline{F}$ is determined by the formula

$$\overline{F} = \frac{1}{2} E_z^2 grad(\varepsilon_{33}^T \varepsilon_0) .$$

On the interface of polar DE in an electric field there are mechanical forces. They are caused by interlaminar Maxwell-Wagner polarization. Mechanical forces arise where dielectric permeability changes abruptly and where electric fields are concentrated (at structural inhomogeneities). The concentration of mechanical and electric fields at the interface of two different DE with a *rigid* inclusion has been analytically calculated by conformal imaging and analytic extension. [Yemets Y.P., Obnosov Y.P., Onofriychuk Y.P. Electric forces on the interface of dielectric media at cylindrical circular inclusion. *J. Appl. Mech. Techn. Phys.* 1993. Vol. 34. No. 4, p. 14-24].

To calculate σ_{mech} (F), it is necessary to use maximum value $E_z^{max} \cong \rangle E_z \langle (H/R)^{1/2}$, where E_z is the averaged value of the field strength, H is the height of the microtip on the electrode surface, R is the radius of its top curvature.

The total value of mechanical stresses will be the sum of two components

$$\sum \sigma_{mech} = \sigma_{mech} + \sigma(\overline{F}).$$

A technique for calculating mechanical stresses $\sigma(\overline{F})$ in pyroelectric crystal plates is given in the Appendix.

At microtips on the electrode surface of a piezoelement the local strength of the electric field in the DE may *exceed the average value by two orders*. As a result, in a relatively weak external electric field electrons are injected into the DE from the cathode, and therefore thin DE layers contacting with metal exhibit semiconductor properties.

A current in the high ohmic DE causes Joule heat, which in its turn increases the concentration of injected electrons. Such a "*swing*" of the electron subsystem raises temperature around micropeaks on the electrode surface and increases the electric instability of the electron subsystem. The physics of the instability may vary, but the common result is a considerably nonlinear dependence of electroconductivity of the dielectric diode on the electric field strength and temperature. In this case there is an S-shape zone in the current-voltage characteristic (CVC). At the first cusp of the function current oscillates in the measuring circuit which contains a dielectric diode [Bogomol'nyi V.M. Theory of nondestructive testing in the electrical degradation of semiconductor devices. *Measurement Techniques*. 2001. Vol. 44. No. 5, p.513-516].

Let us estimate the electric field strength E_{cr}, corresponding to the first appearance of oscillations. The temperature of the dielectric diode ΔT in a constant electric field is determined from the formula

$$\Delta T = I E \tau_t / C_0 \rho, \quad (1)$$

where $I = e \tilde{\mu} n E$ is the electroconductivity current; e, $\tilde{\mu}$ are the electron charge and mobility, E is the electric field strength, τ_t is the electron travel time, C_0 is the DE heat capacity, ρ is the density, n is the concentration of electrons injected into the DE, calculated from the formula

$$n = C_{el} U / e h \Theta, \quad (2)$$

where $C_{el} = \dfrac{\varepsilon_{33}^T \varepsilon_0}{4 \pi h}$ is the capacity, h is the DE thickness; ε_{33}^T, ε_0 are the relative dielectric permeability and dielectric permeability of vacuum in F/m; Θ is the function characterizing the "trapping" of electrons.

E and τ_t are calculated from the formulas

$$E = \frac{U}{h}, \quad \tau_t = \left(\frac{h^2}{\tilde{\mu}U}\right)\Theta. \quad (3)$$

As follows from (1)-(3)

$$\Delta T = \varepsilon_{33}^T \varepsilon_0 \left(\frac{U}{h}\right)^2 \Big/ 4\pi\rho C_0. \quad (4)$$

The field strength $<E>$, averaged against diode thickness, can be determined from (4)

$$\rangle E\langle = \left(\frac{4\Delta T\pi\rho C_0}{\varepsilon_{33}^T \varepsilon_0}\right)^{1/2} \quad (5)$$

For example, for CdS crystals we estimate from (5) at $\Delta T = 1°C$, $\varepsilon_{33}^T = 10$, $C_0 = 0.1$ cal/g·grad., $\rho = 2$ g/cm^3 $E_{ħp} = 3·10^4$ V/cm.

The E_{cr} value from (5) corresponds experimental results. E.g., the maximum value of the electric field strength of a transistor (in the generator mode) is $E = 2.5·10^4$ V/cm (operating frequency $f = 3·10^{11}$ Hz).

On the basis of experimental data, to calculate the concentration of the electric field strength, one can use the Inglis formula for finding mechanical stresses at rigid inclusions (or holes) in the tensile plate

$$E_{cr}^{max} = \rangle E\langle \left(1 + 2\sqrt{\frac{H}{R}}\right), \quad (6)$$

in which H is the micropeak height, R is the radius of its top curvature. It follows from (5) and (6)

$$E_{cr}^{max} = \left(\frac{4\Delta T\pi\rho C_0}{\varepsilon_{33}^T \varepsilon_0}\right)^{1/2}\left(1 + 2\sqrt{\frac{H}{R}}\right). \quad (7)$$

The local temperature of the diode surface can be measured by the differential tomography technique.

The electric field strength is distributed along the thickness of MDM structures unevenly and is determined from the formula

$$E_z(x) = -\frac{3}{2}U\sqrt{x}/L^{3/2}, \quad (8)$$

where U is the voltage on the diode electrodes, L is the dimensionless thickness of the dielectric layer, normalized to a distance unit, i.e. by Debye $\bar{x}$ (marked off the cathode)

$$\bar{x} = \sqrt{\varepsilon_{33}^T \varepsilon_0 kT / 4\pi e^2 n}, \quad (9)$$

where k is the Boltzmann constant, T is the temperature in energy units, e is the electron charge, n is the concentration of injected electrons estimated from the formula

$$n = U\varepsilon_{33}^T \varepsilon_0 / 2\pi h^2 e, \qquad (10)$$

(h is the thickness of the dielectric layer).

Substituting (10) into (9), we obtain $\bar{x} = h(kT/eU)^{1/2}$.

As follows from (8), the electric field strength $E_z(x)$ *at the anode* is the greatest. It is here that the *destruction* of the DE sets in.

If temperature rises (at the first cusp of the S-shape in the current-voltage characteristic) in a constant external electric field, *local low-voltage DE polarization* may take place. When the constant external electric field and spontaneous oscillations of the polarized DE medium occur simultaneously, Joule heating is accompanied by more heating due to energy dissipation caused by *dielectric, mechanical and coupled electromechanic (piezoelectric) losses*. A *negative* electric "resistance" occurs as a result of temperature feedback. When the field strength exceeds the critical point, electric degradation becomes *irreversible*.

3.2. Temperature, tensoresistive and electrostrictive effects in FE

The essence of the tensoresistive effect in FE lies in the changing width of the energetic "prohibited zone" depending on strains.

Displacements change the electron energy level, therefore FE resistance has an essentially nonlinear dependence on strains (tensoresistive effect). Thus, FE and piezosemiconductor-based sensors are 1-2 orders more sensitive than wire tensosensors [9].

Piezosemiconductor tensosensors (1×1.5 mm) measure constant and alternating (at a wide frequency range) pressure and are used in microphones, vibration sensors, accelerometers, etc.

Mechanical stresses and strains, temperature and strength of the controlling electric field cause interconnected changes of polarization and electrochemical potential of FE.

Increasing the sensitivity of MW antennae is one of the main purposes in radiophysics (radiolocation). This is achieved by reducing internal noise in amplifiers and designing new physical methods for receiving electromagnetic waves. *Piezoelectric FE* crystals, which work in the *parametric* oscillation mode, are second in their noise level and signal distortion only to *quantum* amplifiers (lasers) and offer *superior quality* compared to semiconductor diodes. *FE devices* can *electronically tune frequency to millimeter waves*.

The quality of a MDM structure Q may be shown as

$$\frac{1}{Q} = \text{tg}\,\delta_{\text{space}} + \text{tg}\,\delta_{\text{surf}},$$

where the space $\text{tg}\,\delta_{\text{space}}$ and surface $\text{tg}\,\delta_{\text{surf}}$ are the experimentally determined constants, which characterize dielectric, mechanical and piezoelectric losses in the active material.

A MW electromagnetic field penetrates *metallic electrodes* to the depth of the *skin effect*. At the *optimal FE thickness* the required *external energy pumping power* is *minimal*. This condition is fulfilled if energy losses in electrodes are approximately *two times* as high as in FE. Dynamic quality is characterized by the following formula

$$Q_d = \frac{\tilde{m}}{2\omega_c R_S C_0},$$

where $\tilde{m}$ is the depth of capacity modulation, C_0 is the effective *dynamic* capacity, R_S is the equivalent resistance at resonance frequency (characterizing energy dissipation).

For a parametric increase of the amplitude and phase control of MW we can use dielectric nonlinearity of FE in the paraelectric phase, since at temperatures above the Curie point T_C (phase transformation temperature) FE have lower dielectric losses than at $T < T_C$. FE have a unique feature which accounts for their application in radioelectronic and measurement systems, i.e. the quadratic dependence of dielectric permeability on the electric field strength. This leads to a parametric excitation of FE and thus doubles the frequency. Consequently, FE MW amplifiers are superiour to their semiconductor counterparts in electric strength, construction simplicity and manufacturing technology.

Computers are mostly used in information complexes. Due to noise and necessity of transforming analog signals from sensors, their precision and speed of calculation are limited. If high precision is not relevant, one can use optoelectronic devices, in particular for processing two-dimensional images [85-88]. Optical image processing is performed at the speed of light simultaneously for all dots. For example, an optoelectronic system with TV tubes processes an array of 10^5 dots in 0.02 s. A computer which performs 1 million operations per second (using fast Fourier transformation programmes) takes about 10 s for the same data array (not including access time to intermediate memory with 10^5 cells.

Optical analog systems based on transparent FE ceramics (PZLT-type) have less noise and can be controlled only electrically. These optoelectronic devices perform a two-dimensional spectral analysis and simultaneously measure the amplitude and phase signal spectrum [9, 11, 12].

Piezoelectric motors and actuators ensure precise adjustment of the elements of optical analysers [8]. To analyse time signals, we use a scanning light beam in a controlled electric field. A principal feature of FE is that at phase transformation near the Curie point (e.g. at 120°C for BaTiO$_3$) there is a sharp increase in piezo-, pyro and electrooptical coefficients. Thus, the pyroelectric coefficient of barium titanate is 30-40 times greater in a certain temperature range than that of tourmaline. To boost the *sensibility* of *pyroelectric* receivers and pyrovidicons, temperature is specifically raised against the temperature of the environment. Therefore, designing optoelectronic FE elements, one has to *take into account* the *simultaneous* effect of temperature, mechanical stresses and strains, as well as the strength of the controlling electric field.

3.2.1. CVC and mechanical stress-strain state of PZ optical transducers

In relatively weak external electric fields (10-15V) in thin MDM structures charge carriers are injected from the electrodes, and the DE acquires semiconductor properties. Reference describes the effect of emission currents on photoelectric processes in transparent FE ceramics, which is used in optical shutters, spectrum filters, planar optical fibers and holographic memory elements, as well as in optical radiation modulators [Bogomol'nyi V.M., Serebryakov Y.N. Calculation of the electric field in MDM structures (in Russian). *Measurement Techniques*. 1996, No.1 www.wkap.nl/journals/mete].

Let us determine the electric field strength and currents under a constant external field, taking into account electron injection from the cathode. The density of the thermoelectron emission current from the cathode $j(x)$ (x is the coordinate marked off the cathode), the electric field strength $E(x)$ and the concentration of electrons injected into DE $n(x)$ are determined from the following equation system, written in the dimensionless form

$$\frac{dE}{dx} = -n, \quad \frac{dn}{dx} + nE = j(x). \quad (1)$$

The units of the space coordinate $\bar{x}$ (off-cathode distance), the current density $\bar{j}$, the electric field strength $\bar{E}$ and the concentration of injected electrons $\bar{n}$ are calculated from the formulas

$$\bar{x} = \left(\frac{\varepsilon_{33}^{T}\varepsilon_0 kT}{4\pi e^2 n} \right)^{1/2}, \quad \bar{j} = \frac{edn}{\bar{x}}, \quad \bar{E} = \frac{kT}{e\bar{x}}, \quad \bar{n} = N_c e^{\psi}, \quad \psi = \frac{eU}{kT}, (2)$$

where k is the Boltzmann constant, ε_{33}^{T} is the relative dielectric permeability of DE, ε_0 is the dielectric permeability of vacuum, T is the absolute temperature, U is the potential difference on the electrodes of the MDM structure, $d = \mu kT$ (μ is the mobility of free charges), $\bar{x}$ is the Debye screening radius, N_c is the efficient density of quantum states in the conductivity zone, determined by the following expression

$$N_c = 2\left(\frac{2\pi mkT}{h^2} \right)^{3/2},$$

where $\bar{h}$ is the Planck constant, m is the efficient electron mass.

The main system of equations (1) is reduced to one equation

$$\frac{d^2E}{dx^2} + E\frac{dE}{dx} + j(x) = 0. \quad (3)$$

An analytic expression of the properties of the injecting cathode is given in the following edge condition

$$E(0) = 0 \text{ (with } x = 0). \quad (4)$$

To solve equation (3), we also use the following condition

$$\int_0^h E(x)dx = -U, \quad (5)$$

where h is the DE thickness, U is the voltage between the electrodes on the surface of the DE layer with coordinates $z = \pm h/2$.

In accordance with system (1) and edge condition (4), we set the distributions $E(x)$ and $j(x)$ along DE thickness in the following form

$$E(x) = B_1 x^{1/2}, \quad (6)$$

$$j(x) = A_1 + A_2 x^{-3/2}, \quad (7)$$

where A_i ($i = 1,2$) is the unknown constants.

Using (5) and (6), let us calculate B_1 and the function $E(x)$.

$$B_1 = -\frac{3}{2}Uh^{-3/2}, \quad E(x) = -\frac{3}{2}Ux^{1/2}h^{-3/2}. \quad (8)$$

Substituting (6), (7) into equation (3), we obtain

$$-\frac{B_1}{4}x^{-3/2} + \frac{B_1^2}{2} + A_1 + A_2 x^{-3/2} = 0.$$

Putting to zero the total of coefficients at the same powers of x, we find A_1 and A_2; from (7) we obtain the function $j(x)$

$$j(x) = -\frac{9}{8}\frac{U^2}{h^3} - \frac{3}{8}\frac{U}{(hx)^{3/2}}. \quad (9)$$

The first term in formula (9) is a known expression of the current-voltage characteristic, which was obtained without considering the diffusion current of the electrons injected into DE.

Let us view the most common case of piezoelectric transducer geometry and consider a shell of revolution made of electrostrictive ceramics in the non-polar phase (the inverse quadratic piezoeffect), in which a strong constant field induces polarization, whereas a complementary, much weaker electric field creates working strains in the piezoelement.

At thickness polarization the equations of the electrostrictive effect relative to the time-constant mechanical stresses σ_i, strains ε_i and electric field strength E_3 appear as [6]

$$\varepsilon_1 = s_{11}^E \sigma_1 + s_{12}^E \sigma_2 + Q_{12} E_3^2,$$
$$\varepsilon_2 = s_{11}^E \sigma_2 + s_{12}^E \sigma_1 + Q_{12} E_3^2, \quad (10)$$

where ε_i ($i = 1,2$) is the shell strains in the direction of the unit vectors $\vec{\tau}_1$ and $\vec{\tau}_2$; σ_i is the mechanical stresses, s_{1i}^E is the elastic compliances of ceramics, Q_{12} is the electrostrictive constant, E_3 is the electric field strength.

Deducting the correlations of electroelasticity, we accept the following assumptions:

1. The polarization P_3, field strength E_3 and constant Q_{12} linearly depend on the coordinate z, which is calculated from the middle surface of the shell ($-h/2 \le z \le h/2$).

$$P_3 = P_3^0 \alpha, \ E_3 = E_3^0 \alpha, \ Q_{12} = Q_{12}^0 \alpha, \ \alpha = (1/2 + z/h), \quad (11)$$

where P_3^0 and Q_{12}^0 are the values corresponding to the coordinate $z = h/2$, $E_3^0 = -2V_0/h$ (h is the shell thickness);

2. In accordance with the Kirchhoff-Love hypotheses and polarization distribution (11) we write the shell strains in the following form

$$\varepsilon_1 = \varepsilon_i^0\left(1 + \frac{2z}{h}\right) + z\,æ_i, \qquad (12)$$

where ε_i^0 and $æ_i$ ($i = 1,2$) are the relative strains and changes of the main curvatures of the middle surface of the shell;

3. From experimental data we take the dielectric permeability of ferroelectric ceramics as averaged constant along the shell thickness.

From the main equation system (10) it follows

$$\sigma_1 = \frac{1}{s_{11}^E(1-\mu^2)}(\varepsilon_1 + \mu\varepsilon_2 - \hat{E}_3),$$

$$\sigma_2 = \frac{1}{s_{11}^E(1-\mu^2)}(\varepsilon_2 + \mu\varepsilon_1 - \hat{E}_3), \qquad (13)$$

$$\hat{E}_3 = (1+\mu)Q_{12}(E_3^0)^2, \quad \mu = -\frac{s_{12}^E}{s_{11}^E}, \text{ } \mu \text{ is the Poisson's ratio.}$$

Expressing internal mechanical forces in the shell T_1 and T_2 and bending moments M_1 and M_2 through the integrals of stresses σ_1 and σ_2, we obtain the following correlations of electroelasticity

$$T_1 = D_T\left[\varepsilon_1^0 + \mu\varepsilon_2^0 - \frac{(1+\mu)}{4}Q_{12}^0(E_3^0)^2\right],$$

$$T_2 = D_T\left[\varepsilon_2^0 + \mu\varepsilon_1^0 - \frac{(1+\mu)}{4}Q_{12}^0(E_3^0)^2\right], \qquad (14)$$

$$M_1 = D_M[(2/h)(\varepsilon_2^0 + \mu\varepsilon_1^0) + æ\,1^{+\mu} \text{ } æ\,e\,2 - E_{3M}],$$

$$M_2 = D_M[(2/h)(\varepsilon_2^0 + \mu\varepsilon_1^0) + æ_2 + \mu\,æ_1 - E_{3M}],$$

$$D_T = \frac{h}{S_{11}^E(1-\mu^2)}, \qquad\qquad D_M = \frac{h^3}{12S_{11}^E(1-\mu^2)},$$

$$E_{3M} = \frac{3}{4}\frac{(1+\mu)}{h}Q_{12}^0(E_3^0)^2.$$

Let us consider a console-mounted homogeneous piezoelectric plate with thickness polarization. We can determine the free end deflection under the following conditions, which are met along the whole length

$$T_1 = M_1 = 0, \text{ } æ_1 = \text{const}, \qquad (15)$$

$$(T_2 = M_2 = \varepsilon_2 = æ_2 = \mu = 0).$$

From the first and third equations in (14) and (15) it follows that

$$\varepsilon_1^0 = Q_{12}^0 (E_3^0)^2 / 4, \; (2/h)\varepsilon_1^0 + \mathbf{x}_1 - E_{3M} = 0. \qquad (16)$$

From equation system (16) we determine the curvature change $\mathbf{x}_1$.

$$\mathbf{x}_1 = Q_{12}^0 (E_3^0)^2 / 2h \cdot$$

In view of the latter equation and (12) the deflection of the free end of the plate can be calculated by the following formula

$$f = \frac{l^2 \Delta \mathbf{x}_1}{2} = \frac{l^2 Q_{12}^0 (E_3^0)^2}{2h}, \qquad (17)$$

$$\Delta \mathbf{x}_1 = [\varepsilon_1(h/2) - \varepsilon_1(-h/2)]/h = Q_{12}^0 (E_3^0)^2 / h,$$

where l is the plate length, $\varepsilon_1(\pm h/2)$ are the relative strains on plate surfaces with coordinates $z = \pm h/2$.

With the field strength $E_3^0 = 6000$ V/cm and $l = 30$ mm; $h = 0.15$ mm; $Q_{12}^0 = -1.45 \times 10^{-10}$ mm^2/V and $\mu = 0.34$, the free end deflection $f = 0.14$ mm.

Measuring how thickness polarization is distributed in thin-layer piezoelectric structures presents certain technical problems. Formula (17) can be used for obtaining an integral assessment of changing polarization in the thickness of thin piezoelectric layers (10-200 μm). Formula (17) may help to determine the deflection of the free end of a console-mounted sample. Comparing the result of the calculation with experiments helps to determine by design-experiment the distribution of thickness polarization.

When transparent FE semiconductors are subjected to an electric field, they develop piezoelectric strains, which cause *the mechanooptical Kern-Harbeke effect*. Electrooptical properties of FE *largely depend on mechanical stresses*. For example, at σ = 15 *MPa* the electrooptical coefficient of a Cd_2NbO_7 crystal changes by 50%.

Changing only the electric field is the simplest way to control *electrooptical and piezoelectric* properties of FE semiconductors, which can be used in various optoelectronic devices with *piezo- and elastooptical effects* in play at the same time.

3.3. Automatic thermostabilization of FE resonators (tandels, thermostats)

In the TANDEL (thermostat), the construction of which is well known, internal thermostabilization occurs due to the nonlinear dependence of *di-*

electric losses on the temperature of FE at phase transformation [1]. A similar effect was discovered in FE resonators excited by MW moderate-power generators and was subsequently used in *microthermostats* for VLSI, heat radiators, MW phase commutators and IR optoelectronic devices.

Semiconductor optoelectronic devices (light-emitting diodes and photo-diodes) have a *low temperature stability*. For example, the power of IR light-emitting diodes and planar optical fibers at ~100°C is reduced by 2-3 times (as against +25°C). The temperature coefficient of the change in photoresistor parameters is (0.2-2%)/°C, and (0,1÷0,5)%/°C for photodi-odes.

The use of resonance devices for thermal stabilization depends on whether their working temperature can be controlled by the excitation frequency. The working temperature of TANDELs is determined by the phase transformation temperature of polar DE; thus, is not changed during use. This accounts for the superiority of frequency-controlled FE MW resonators.

3.4. Choosing the optimal excitation frequency for PE

At harmonic oscillations the heat conductivity equation for a PE takes the form of [Bogomol'nyi V.M. Calculation of energy dissipation and heating temperature in piezoelectric MDM structures. *Measurement Techniques.* 1998. No. 8. www.wkap.nl/journals/mete]

$$\rho C \frac{\partial T}{\partial t} - \lambda \frac{\partial^2 T}{\partial t^2} = \omega \sum_{n=1}^{3} \Phi_i \, (i = 1,2,3), \qquad (1)$$

where T is the temperature that arises at resonance heating, t is the time, λ is the heat conductivity coefficient, C is the heat capacity; Φ_1, Φ_2, Φ_3 are the function components of heat emission due to dielectric, mechanical and piezoelectric losses, calculated from the formulas

$$\Phi_1 = C_{el} V_0^2 \, \text{tg}\, \delta_{de} / v , \qquad (2)$$

$$\Phi_2 = \frac{E_{elast} \, \text{tg}\, \delta_{mech}}{2(1+v)} \left[\varepsilon_1^2 + \varepsilon_2^2 + \varepsilon_3^2 + \frac{v(\varepsilon_1 + \varepsilon_2 + \varepsilon_3)^2}{1-2v} \right], \qquad (3)$$

$$\Phi_3 = -E_z \, \text{tg}\, \delta_{pe} [e_{33}\varepsilon_3 + e_{31}(\varepsilon_1 + \varepsilon_2)], \qquad (4)$$

where C_{el} is the electric PE capacity, v is the PE volume, E_{elast} is the Young's modulus; e_{33}, e_{31} are the piezoelectric constants; $\text{tg}\, \delta_{de}$,

$tg\,\delta_{mech}$, $tg\,\delta_{pe}$ are the tangents of the angles of dielectric, mechanical and piezoelectric losses, ε_i ($i = 1,2,3$) are the relative deformations, $V = \pm V_0$ are the electric potentials of the PE electrodes.

The edge conditions for the heat conductivity equation are

$$T\big|_{t=0} = T_0, \quad \lambda\frac{\partial\partial T}{\partial z}\bigg|_{z=\pm\frac{h}{2}} = \mp\alpha(T - T_0), \qquad (5)$$

where T_0 is the initial temperature, α is the coefficient of heat emission, T_0 is the temperature of the environment.

f, kHz	$W\cdot 10^4$ [m]	$\Phi_{de}\cdot 10^{-3}$ [W/m³]	$\Phi_{mech}\cdot 10^{-3}$ [W/m³]	$\Phi_{pe}\cdot 10^{-3}$ [W/m³]	ΔT, K
30.4	0.156	0.677	8.831	13.60	1.235
30.8	0.217	0.686	16.970	16.75	1.84
32.8	-0.227	0.730	20.430	-5.319	0.93
33.2	-0.156	0.739	9.976	-1.49	0.51

Table 6. Dependence of heating temperature on frequency

Expression (4) should include the phase shift between the external electric field and piezoelement deformations.

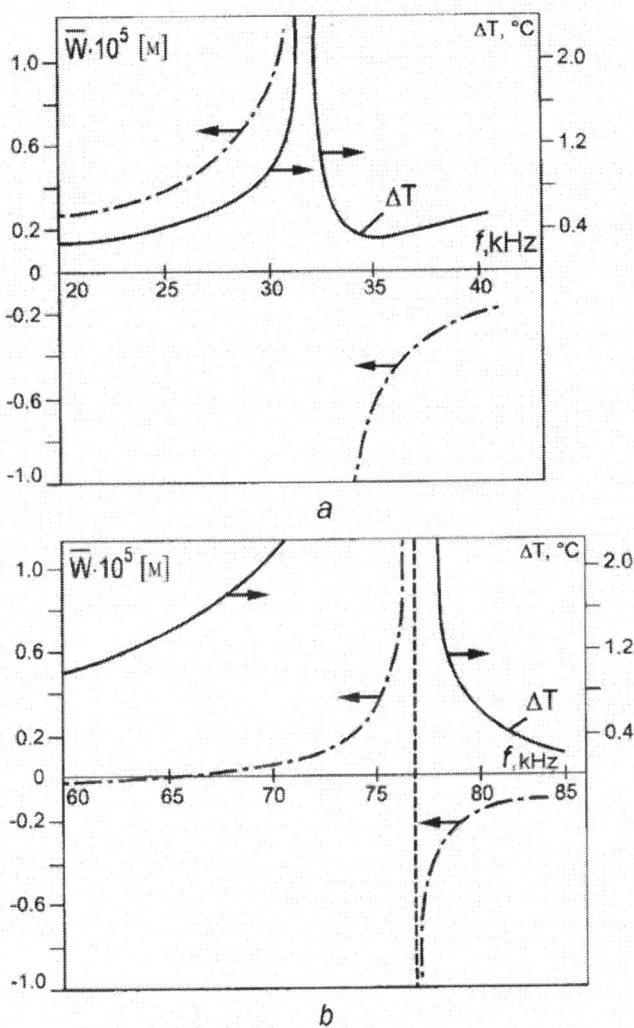

Fig. 12. Curves of heating temperature and deflections of a cylindrical piezotransducer; a is the first lowest resonance frequency, b is the second resonance frequency

Φ_1 and Φ_2 functions are quadratic (because they depend on V_0^2 and ε_i^2 ($i = 1,2,3$)) and *always positive*; Φ_3 *function* on the right-hand branch of the resonance curve has *a negative value* (see (4)).

Φ_3 function (characterizing piezoelectric losses) is positive before the resonance frequency and *negative on the right-hand branches* of resonance curves. At resonance frequency $\Phi_3 = 0$, since the phase shift angle between the electric field and mechanical strains of the piezoelement is 90°. We have calculated Φ_i functions ($i = 1,2,3$) and numerically solved the heat conductivity equation for a piezoceramic cylindrical shell made of barium lead zirconate titanate ceramic (with $z_0 = 14$ mm, $h = 2$ mm, $V_0 = 12.4$ V). The results are given in Table 6. The curves of shell-length-averaged *stationary heating temperature* ΔT and *displacements* (deflections) *W*, normal to the middle surface of the shell, are shown in Fig. 12. At the same strain amplitude, if $f_1 = 30.4$ kHz, ΔT equals 1.235 °C; if $f_2 = 33.2$ kHz, ΔT equals 0.51 °C (on the right-hand branch of the first resonance curve) we observe a local minimum of heating temperature (see Table 6).

As our calculation and experiments show, the excitation frequency at which the electromechanical connection has the highest value, equals $(0.8 - 0.95)\omega_a$ (see Fig. 12) [13].

The oscillation amplitude of crystal quartz near resonance frequency grows (*in proportion to mechanical quality, by 10^4 times*). At the same time, *polarization* also reaches its maximum. In the electric circuit which connects the electrodes of the quartz resonator, the electric current increases similarly. Due to this a quartz resonator is used to stabilize the frequency of time devices and to filter oscillating circuits (*the principal element* of the mobile radiotelephone).

On the right-hand branch of the resonance curve, when the strength of the internal electric field decreases and the displacement current increases (in proportion to oscillation frequency), the current in the external circuit and the biasing current in the piezoelement are *directed at each other*, which corresponds to the minimum of the total current and energy consumed (see Fig. 13) (or the maximal susceptibility of the piezoelectric).

The piezoelement has an eigenfrequency of mechanical oscillations, which depends on capacity, inductivity and electric resistance. It is a part of the "external voltage source – polar dielectric " electric circuit with its *eigenfrequency of the electric resonance*. For a stronger *electromechanical coupling, the frequencies of mechanical and electric resonances should coincide*.

Traditionally, the piezomodule d_{ij} for piezoceramic materials in the dynamic mode for various oscillation modes is measured by the relation of

d_{ij} and the coefficient of electromechanical coupling k_{ij}. The piezomodule is determined by measuring mechanical resonance frequencies ω_r and antiresonance frequencies ω_a, as well as the capacity of the piezoceramic element at low frequency C^T. This is the so-called "resonance-antiresonance" method.

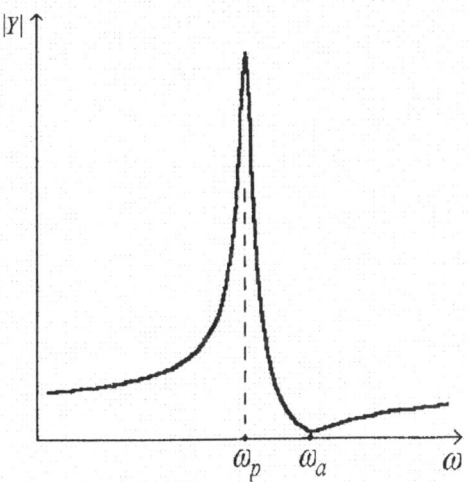

Fig. 13. Frequency characteristic of conductivity in the resonance zone of a piezoceramic element

The values of ω_p and ω_a are experimentally determined by the frequency characteristic of the conductivity modulus $|Y(\omega)|$. A typical relation $|Y(\omega)|$ is shown in Fig. 13 [2].

At parametric oscillations, the optimal mode of the electric excitation of PE transducers is achieved when the frequency of the energy source is *two times* higher *than the eigenfrequency* of the piezoelement's *mechanical oscillations* (determined only by the elasticity modulus, size and shape of the transducer), or at *shock-and-vibration resonance*.

In the *shock-and-vibration resonance* mode the electric circuit of the excitation source generates impulses lasting approximately 1/8 of the oscillation period of the piezoelement. At each half-wave of the oscillations the direction of the electric impulse of excitation corresponds to the direction of mechanical oscillations. In the beginning of each oscillation wave the electromechanical system receives a shock impulse, after which the piezoelement oscillates at the eigenfrequency of elastic mechanical oscillations. The excitation system in a telemetric system works similarly in the mode of relaxation self-oscillations.

3.5. Synchronization of oscillations in electromechanical transducers

Self-organisation, discovered in the early 17th century by Christian Huygens, is one of the first instances of autosynchronization in quasilinear mechanical systems. Today it is relevant for the design of computing devices and automatic control and measurement systems [63, 82, 84, 97, 105-107].

If a nonlinear autooscillating dynamic system with the eigenfrequency ω_0 is subjected to an external "force" with the frequency ω, at small values of $(\omega - \omega_0)$, it does not *swing*. On the contrary, unlike elastic systems, it *self-synchronizes* at the *recombination resonance frequency* (due to energy dissipation). This phenomenon was mathematically expressed in the fact that the Rayleigh and van der Pol nonlinear differential equations have periodic solutions resistant to fluctuations (it is represented by *limit cycles*, independent of initial conditions).

The effect of fluctuations (inevitable in real life) on autooscillations in measurement systems has practical importance for *time standards* (generators with a high stability of frequency), devices measuring time intervals and *separating weak signals from "white" and "colour" noise*.

In 1954 J. von Neumann demonstrated how an element of a nonlinear dynamic system amplifies, switches and memorizes information.

In computer and measurement technology the most promising are solid-state electronic devices with a negative differential conductivity, whose current-voltage characteristics (CVC) are N-shape [71]. These measurement devices are characterized by *self-synchronization of the sensor and driving circuit frequencies* [Bogomol'nyi V.M. Use of electric instability effect for non-destructive testing of MDM structures. *Measurement Techniques*. 1998. No. 8. www.wkap.nl/journals/mete]

Self-synchronization occurs in various natural objects and technical systems, "which automatically develop a common rhythm of co-existence *despite different individual rhythms and extremely weak interconnections*".

Autosynchronization can only happen when the susceptibility of the dynamic system to fluctuation intensity is nonlinear and has a maximum (e.g., CVC should have an extremum, i.e. cusp). As shown by experimental and theoretical study, *the structure of the electromagnetic excitation signal may be arbitrary*. The signal can be *impulse, quasiperiodic* and even *noise*, i.e. contain a spectrum of various frequencies. This chapter demonstrates how tunnel diodes in close range telemetry (contactless information transfer from fast-rotating objects) can be effectively replaced with *permeable base submicron transistors (PBST)*, which can *simultane-*

ously function as *ultrasensitive sensors* and *autooscillation generators*. Like tunnel diodes, they have an N-shape CVC, and can be switched with an energy of about *one femtojoule*.

Resonance transducers of electric energy into mechanical energy, based on polar DE, piezosemiconductor and magnetostrictive materials, are used as a *precise tool* for the non-destructive testing of materials and construction elements. Various resonance transducers have common laws of autosynchronization, which can be explained by the theory of nonlinear dynamic systems [90, 91, 94, 103, 104].

Autooscillations can be excited with a periodic succession of coherent impulses (or with relaxation oscillations). This chapter states the conditions for the *phase* autosynchronization of piezoelectric transducers.

An example of autosynchronization is the electrodynamic system for the distance measurement of parameters of fast-rotating objects: turbines in power equipment and motors, centrifuges with $10^5 - 2*10^5$ rotations/min [Seryoznov A.N. et al. Negatronics. Novosibirsk. Nauka, 1995]. These systems gauge the temperature of the moving object, its stresses and strains in its elements, and in some cases the orientation and size of defects of its inner structure.

Autowave synchronization in piezosemiconductors and polar dielectrics can be used for superfast analog information processing in real time (including optical information) [6].

The simplest and at the same time most precise are the informational measurement systems that contain resonance circuits of various physics (with frequency measured in a contactless way). In spite of the differences between the electromechanical and piezoelectric systems, they have common features: 1) the resonance response of the "passive" (excited) element, 2) the "active" low-quality element working at an *"antiresonance"* frequency, 3) the choice of the optimal frequency of electric excitation on the right-hand branches of resonance curves.

This paragraph shows how two oscillation circuits (the "passive", resonance or sensor-containing, and the "active" circuit, the source of electromagnetic excitation) can be matched by the phase of electromagnetic field components. This problem was studied experimentally and theoretically in [Sambyrski A.I., Novik V.K. Contactless measurements of parameters of rotating objects. Moscow, Mashinostroyenie, 1976]. This book gives experimental proof to P.N.Lebedev's theories of the early 20th century (see Fig.14, cf. Fig.15, 16) [Lebedev P.N. Experimental study of the ponderomotive effect of electromagnetic waves on resonators. Coll. Papers. Moscow, Ac.Sci., 1963].

Experimental data on the conditions for autosynchronization of the oscillation excitation circuit and frequency in a PE electromechanical system; temperature behaviour analysis of piezoelectric transducers which transform electric energy into mechanical; analysis of changes in the internal electric field strength at resonance and anti-resonance frequencies can be found in [Bogomol'nyi V.M. Use of electric instability effect for nondestructive testing of MDM structures. Measurement Techniques. 1998. No.8. www.wkap.nl/journals/mete]. Having studied the resonance interactions of two electric LC-circuits, L.I.Mandelstamm and N.D.Papaleksi built a *precision measuring device for frequencies* and damping factors (their modern counterparts measure the oscillation period accurate to 10^{-9} s) [Papaleksi N.D. Evolution of the concept of resonance. Collect. Papers. Ed. S.M.Rytov. Moscow: Ac. Sci. 1948].

Quantitative evaluation and understanding of the electrodynamic processes of energy dissipation and heating in polar DE due to dielectric, mechanical and *coupled electromechanic (piezoelectric)* losses near resonance is necessary to design precision *frequency measurement devices* (including remote sensing), highly selective radio and MW filters, temperature autostabilization devices (TANDELs, ferroelectric resonators), temperature-stabilized *acoustoelectronic generators*, *electronic devices and time standards*, as well as frequency, current and voltage piezotransformers [1, 2, 23, 24].

The automatic control theory, as well as radioelectronics and optoelectronics, study *autowave phenomena* in piezo- and ferroelectrics, which can be used for identifying useful signals in noise, in information recording devices based on computing structures (similar to perceptrons) and in memory elements of microprocessors [9].

Physically, nonlinear ferroelectrics have a dynamic memory and can memorize oscillation phases, amplify electromagnetic signals, filter undesirable higher harmonics, identify image contours and regenerate lost recording elements. All these solid-state electronic devices are characterized by the *autosynchronization of oscillation processes* in piezoelectric media with energy dissipation [45].

Autosynchronizing generators in an electromechanical close range telemetry system ensured a stable remote transmission of information from fast-rotating objects (to $n = 10^5$ rotations/minute). The most precise and at the same time simplest were the measurement systems with *a frequency output*. The *electromechanical* coupling of the resonance circuits sets the *principle precision limit* of telemetry.

A one-channel telemetric system consists of an autogenerator of relaxation oscillations (the "active" element) and an LC-circuit (the "passive" resonance element). The autogenerator contains transistor and *negatronic* multivibrators. The "passive" element has high-quality oscillation circuits, including sensors. The "active" element is of low quality. The working point of the tunnel diode corresponds to the second cusp of the N-shape CVC [Sambyrski A.I., Novik V.K. Contactless measurements of parameters of rotating objects. Moscow, Mashinostroyenie, 1976].

This paragraph contains a formula for calculating critical strength, under which autooscillations of the acoustoelectric generator occur. It is also shown that the results of P.N.Lebedev's experiments correspond to contemporary scientific views. One of the results was thus formulated: "The mechanical effect of the exciting electromagnetic wave on the resonator depends only on their frequencies". Industrial telemetric systems have proved to be the simplest and most precise. Experiments gave the most rational correlations of partial frequencies $(\omega_2 / \omega_1) < 1$ (where ω_1 is the frequency of the vibrator, and ω_2 is that of the resonator) [Sambyrski A.I., Novik V.K. Contactless measurements of parameters of rotating objects. Moscow, Mashinostroyenie, 1976].

P.N.Lebedev studied the interaction of a passive electromechanical system ("condenser") with an external source of electromagnetic excitation. Subsequently, A.Sommerfeld established that *before resonance* a considerable supply of external energy causes small changes in the frequency of the electromagnetic excitation source ("*vibrator*"). After the maximum of the oscillation amplitude (the peak of the resonance curve) there is a *sharp* change in excitation frequency. We are speaking about the inertial response of the oscillation contour in the electromechanical "passive" element on the oscillation frequency of the "external driving force".

The interaction of the oscillation circuits of various transducers with an external power source has further been studied in mechanics, electrotechnics and electronics [63, 67, 82, 89, 99, 102-107].

The theory of nonlinear oscillations explains "*self-synchronization*" or "*frequency capture*", characterized by *stability* towards fluctuations and *independence* of *initial conditions*. In each element of this autowave system, energy is "pumped in" by an external source and "drained", or dissipated. The interaction of fast electronic processes and "slow" polarization relaxation processes in a crystal lattice causes ferroelectric instability and sustained autooscillations.

The principal feature of "*frequency capture*" is that the autosynchronization of various oscillations (mechanical, electric and *coupled electromechanic, or piezoelectric*) is caused by energy losses themselves (the so-called "inner friction" in non-conservative mechanical systems).

A nonlinear dynamic system near resonance due to energy dissipation changes some of its properties in different ways: smoothly, quickly or stepwise (the vector directions of the electric field strength and current, phase shift between them). The system "self-adapts" to the external load. As a result, the frequency of the transducer autosynchronizes with the frequency of the external load. When piezoelectrics oscillate, due to the *direct piezoeffect* they "yield" *internal energy*, which returns to the external energy source (i.e. the transducer and the permissible maintain a constant energy exchange). The combination of "slow" (polarization) and "high speed" movement changes (in the electronic subsystem) leads to the "frequency capture".

At resonance oscillations of PZ transducers (at inverse piezoeffect) there is an *unusual increase* (by 4-20 times) in the strength of *the "excited" internal electric field* compared to the original electric field from the *power supply*. The growth can be explained by the combination of the "direct" and "inverse" piezoeffects, as well as by the acoustoelectronic generation of a supplementary electric charge at the boundary where the metal electrode contacts with the resonating piezoelectric.

P.N.Lebedev and A.Sommerfeld's experiments showed that an electromechanical system before resonance "takes" the *maximum* of energy from the external source, and *minimum* of energy after resonance. This result is confirmed by later research. It has been found that in PZT-4 ceramics the *optimal excitation frequency* is on the *right-hand branch of the resonance curve* (at a small distance from antiresonance frequency peak). This corresponds to the minimum of electric external energy consumed by the piezotransducer, and the maximum *dielectric "susceptibility" of piezoceramics* [13].

The changes that happen near resonance frequency *inside* piezoelectric *transducers* and in sources of electric excitation, are vital in the design of *sensors*. Quantitative evaluation and understanding of electrodynamic processes is problematic, since the *integral characteristics* of piezotransducers (capacity, inductivity and resistance in the external electric circuit) are *the only reliable source of information about inner changes in oscillating polar DE*.

Oscillations in DE cause *the displacement current* (due to a change in *polarization*). In thin-film MDM structures near micropeaks on the electrode surface we observe thermo- and autoelectronic *injection currents. Thus, polarization and electroconductivity are difficult to separate.*

After the peak of the resonance curve, when the strength of the internal electric field lessens, *the biasing current* (or capacity current) and *the electroconductivity current* are directed *towards each other*. This explains the minimum of summary current and consumed external energy in the interval between resonance and antiresonance frequencies. This condition is characteristic of various oscillation processes (Fig. 13).

Under an external constant electric field in MDM structures current autooscillations occur at the *initial reversible stage of the electric aging of a DE*, if the times of the electron energy relaxation and dielectric relaxation are *equal* [Bogomol'nyi V.M. Theory of nondestructive testing in the electrical degradation of semiconductor devices. *Measurement Techniques*. 2001. Vol. 44. No. 5, p.513-516]. A quantitative assessment of ponderomotive (mechanical) forces in an electromechanical resonator was obtained by P.N.Lebedev, who wrote: "If we employ the word "movement" to denote the magnetic and electric state of the *medium polarization*, determined by *exciting* and *excited* oscillations, the dK impulse will stand for the attracting force, since we consider two *polarizations* like-directed if their field lines form a sharp angle. The driving mechanical forces will then correspond to the attraction... We shall obtain the same results if instead of polarization consider the phenomenon of induction and changes in the electric charges of a resonator." [Lebedev P.N. Experimental study of the ponderomotive effect of electromagnetic waves on resonators. Coll. Papers. Moscow, Ac.Sci., 1963] (Fig. 14). This explanation of the ponderomotive effect that electromagnetic waves have on electromechanical resonators *faithfully reflects* the dynamics of coupled resonance electronic and polarization processes in piezoelectric MDM structures.

In P.N.Lebedev's experiments "the exciting movement" was created by a solenoid, a source of an alternating electromagnetic field. The resonator was a "*condenser*" (in the shape of non-solid cylindrical quadrants made of thin metal wires). The solenoid and resonator were coaxially mounted on the same quartz filament, which performed torsional oscillations under harmonic electric excitation. Electric forces, affecting the charges accumulated on the condenser quadrants (according to the Coulomb's law), excited electromechanical oscillations. The source of electromechanical oscillations, the "vibrator", had a low quality factor Q, and the resonator was of high quality. Similarly, in a telemetric system [Sambyrski A.I., Novik V.K. Contactless measurements of parameters of rotating objects. Moscow, Mashinostroyenie, 1976] the oscillation circuit of external electric excitation (the "vibrator") was low Q, and the oscillation LC-circuit (the "resonator", connected with a sensor on the rotating object), was high Q.

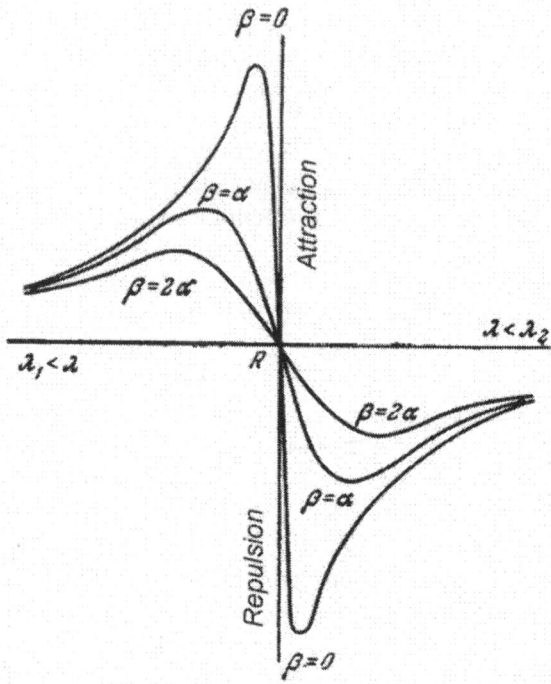

Fig. 14. The curves of mechanical (ponderomotive) forces in the "condenser" [Lebedev P.N. Experimental study of the ponderomotive effect of electromagnetic waves on resonators. Coll. Papers. Moscow, Ac.Sci., 1963]. *(α, β are the constants characterizing internal losses in an electromechanical system, λ is the wavelengths)*

Figure 14 illustrates a change in the mechanical force acting on the resonator when exciting oscillation is constant, and the period and attenuation of the resonator alternate (depending on oscillation frequency) [36]. The exciting "movement" $f(t)$ is an attenuating sinusoid oscillation [36]

$$f(t) = a_0 e^{-\alpha t} \cos a t, \quad (1)$$

where a_0 is the initial amplitude of oscillations, a is the circular angular frequency (rad/s), α is the attenuation (experimentally studied in [38-40]). The oscillation of the resonator consists of forced and free (determined by the resonator's eigenfrequency) oscillations and has the form of [36]

$$\varphi(t) = A e^{-\alpha t} \sin(at + a') + B e^{-\beta t} \sin(bt + b'), \quad (2)$$

where A, B are the amplitudes; a', b' are the initial phases of the oscillations, β is the coefficient which characterizes losses.

The excitation impulse dK for a mechanical system is proportionate to the product of the driving and excited movements.

$$dK = f(t)\varphi(t)dt . \quad (3)$$

The value K, which is in direct proportion to the ponderomotive force, is plotted in Fig. 14 on the ordinate axis and calculated from formula [36]

$$K = E\frac{a(b-a)}{(\alpha+\beta)^2 + (b-a)^2}, \quad (4)$$

where E is the amount of energy of the electromagnetic field affecting the resonator. The abscissa axis in Fig. 14 shows the wavelengths in the resonator, characterizing the change in oscillation frequency.

The function $K(\omega)$ in Fig. 14 coincides in form with the function of piezoelectric losses (or coupled electromechanical losses in a piezoceramic transducer), measured in [Mezheritsky A.V. Energy losses in piezoceramics at electric excitation. *Electricity*. 1984. No. 10, p.65-67(in Russian, translated into Eng.)] Fig. 15.

The heating temperature of the piezoceramic shell was calculated from the equation of heat conductivity (at stationary oscillation)

$$\lambda\frac{\partial^2 T}{\partial t^2} = \omega\sum_{i=1}^{3}\Phi_i (i = 1,2,3), \quad (5)$$

where λ is the heat conductivity coefficient, ω is the circular oscillation frequency, Φ_i is the components of the heat emission due to dielectric, mechanical and piezoelectric losses, calculated from the formulas

$$\Phi_1 = C_{el}V_0^2\, tg\,\delta_{de}/v, \quad (6)$$

$$\Phi_2 = \frac{E_{elast}\, tg\,\delta_{mech}}{2(1+\mu)}\left[\varepsilon_1^2 + \varepsilon_2^2 + \varepsilon_3^2 + \frac{\mu(\varepsilon_1 + \varepsilon_2 + \varepsilon_3)}{1-2\mu}\right], \quad (7)$$

$$\Phi_3 = -E_z\, tg\,\delta_{pe}\left[e_{33}\varepsilon_3 + e_{31}(\varepsilon_1 + \varepsilon_2)\right], \quad (8)$$

where C_{el} is the electric capacity of the piezotransducer (PT), $V = \pm V_0$ is the amplitude of the electric potentials on the PT electrodes; $tg\,\delta_{de}$, $tg\,\delta_{mech}$, $tg\,\delta_{pe}$ are the tangents of the angles of dielectric, mechanical and piezoelectric losses, E_{elast} is the elasticity modulus for piezoceramics, e_{ij} ($i = 1,2,3$) are the piezoelectric constants, E_z is the electric field strength, ε_1, ε_2, ε_3 are the relative strains.

Functions Φ_1 and Φ_2 are *quadratic* and maximal at resonance frequency; thus, they are always *positive*. Function Φ_3 (characterizing electromechanical losses) is *linear* and *negative* on the right-hand branch of the resonance curve (see Fig. 15).

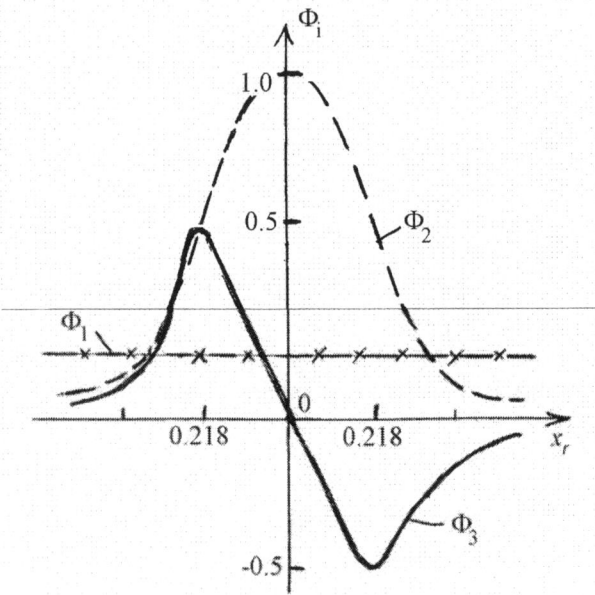

Fig. 15. The dependence of dielectric, mechanical and coupled electromechanical (piezoelectric) energy losses on the frequency of off-tuning from resonance – X_r [Mezheritsky A.V. Energy losses in piezoceramics at electric excitation. Electricity. 1984. No. 10, p.65-67(in Russian, translated into Eng.)]

Frequency-based changes in reactive resistance X_S of a piezoelectric transducer are shown in Fig. 16 [2]. Reactive resistance is proportionate to ponderomotive forces and the function of electromechanical losses Φ_3 (at resonance f_s and antiresonance frequencies f_p function X_S is zero). Stable autooscillations occur around frequency f_{po} (at R > 0) on the right slope of the function X_S ($\omega_s < \omega_{po} < \omega_p$) [2].

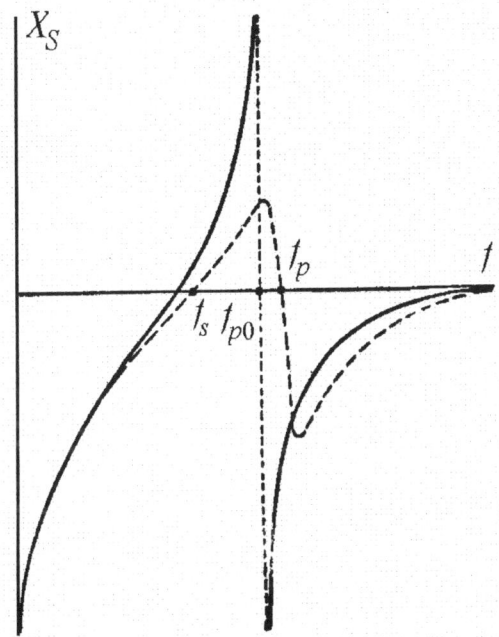

Fig. 16. Change in reactive resistance of a piezoelectric resonator X_s (for the ideal crystal). The dotted line is related to $R > 0$ (real situation). The solid line corresponds to $R = 0$ (R is active resistance), f_s is the frequency of serial resonance, f_p is the frequency of parallel resonance (antiresonance frequency) [2]

It seems difficult to ensure stable amplitudes of the electric field strength (and current) in oscillating piezoelectrics, as well as the *phase shift* between them. Taking into account electron injection from the cathode into PE, under certain conditions it is possible to keep the internal electric field strength in acoustoelectronic generators *independent of the frequency of the external electric excitation* [Bogomol'nyi V.M. Calculation of piezoceramic transducers at harmonic excitation. *Measurement Techniques.* 1994. No.12 www.wkap.nl/journals/mete]. There is a frequency at which the electric field strength in an MDM structure remains constant when ω_c / ω_a changes (see Fig. 17). In Fig. 17 the curves correspond to: $1 - \omega_c / \omega_a = 10$ ($\omega_c = \dfrac{\sigma}{\varepsilon_{33}}$ is the frequency of *dielectric relaxation*, σ is the conductivity, ε_{33} is the dielectric permeability; ω_a is the antiresonance frequency), $2 - \omega_c / \omega_a = 1$; $3 - \omega_c / \omega_a = 0.1$; $4 - $

$\omega_c / \omega_a = 0$. $Q = 50$ is the quality of the piezoelectric. Fig. 17 a) corresponds to $K^2 = 0.1$, 17 b) to $K^2 = 0.025$. $K = \dfrac{e^2}{\varepsilon_{33}\tilde{c}}$ is the coefficient of electromechanical coupling (CEMC) (where e is the piezoelectric constant, $\tilde{c}$ is the elastic constant characterizing the "rigidity" of the piezoelectric). As demonstrated in the graph of the normalized electric field strength (in relative units), all curves meet approximately at one point, which corresponds to the frequency range $0.975\omega_a < \omega < 0.99\omega_a$. Curve 1, the graph of the E_p functions ($\omega_c / \omega_a = 10$) remains virtually constant at changing frequencies.

Choosing a piezoelectric with σ and ε_{33}^T *corresponding to the relation* $\omega_c \approx 10\omega_a$, one can gain a stable electric field strength. This leaves *only one parameter* in the electric source circuit, the phase shift between the current and voltage (cos φ), which can be controlled with a certain precision, thus ensuring autosynchronization in the simplest way.

Let us estimate the electric field strength of the MDM structure E_{cr}, under which at the initial reversible stage of electric breakdown appears an N-shape CVC (a thin layer of a polar dielectric contacting with a metal electrode has an effect of unipolarity and CVC functions like an n-p transition). The frequency of the dielectric relaxation ω_c is determined by the formula

$$\omega_c = \frac{\sigma}{\varepsilon_{33}}, \quad (9)$$

where $\varepsilon_{33} = \varepsilon_{33}^T \varepsilon_0$ is the dielectric permeability, σ is the specific electroconductivity, determined by the expression

$$\sigma = e\tilde{\mu}n, \quad (10)$$

where e is the electron charge, $\tilde{\mu}$ is its mobility, n is the concentration of electrons injected into DE, calculated from the formula

$$n = \frac{E_z \varepsilon_{33}}{he}, \quad (11)$$

where E_z is the electric field strength, h is the DE thickness.

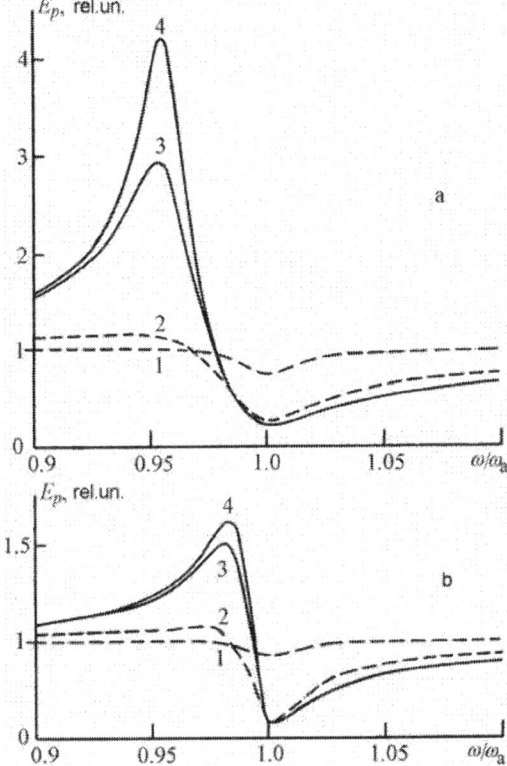

Fig. 17. Dependence of the internal electric field strength at the piezoelectric-electrode contact surface. a) corresponds to $K^2 = 0.1$; b) corresponds to $K^2 = 0.025$ (K is the CEMC) [Bogomol'nyi V.M. Calculation of piezoceramic transducers at harmonic excitation. *Measurement Techniques*. 1994. No.12 www.wkap.nl/journals/mete]. Note that these curves correspond to P.N.Lebedev's experiments.

The antiresonance frequency of the DE layer at thickness oscillations can be calculated by the formula

$$\omega_a = \frac{\pi}{h} \sqrt{\frac{E_{elast}}{\rho}}, \qquad (12)$$

where E_{elast} is the dynamic modulus of DE elasticity, ρ is the DE density. Taking into account the data given in Fig. 17, if $\omega_c \approx 10\omega_a$, from (9-12) we obtain the following formula

$$E_{cr} = \frac{10\pi}{\widetilde{\mu}} \sqrt{\frac{E_{elast}}{\rho}}.$$

For gallium arsenide, taking $E_{elast} = 1.2 \cdot 10^5$ MPa, $\rho = 5.2$ kg/m^3, $\tilde{\mu} = 0.81$ m^2/V·s, we obtain $E_{cr} = 1.89 \cdot 10^5$ V/m.

The given formulas (1)-(5) can be used for calculating the working mode parameters of a PBST, which is preferable to a tunnel diode in a telemetry system (because it consumes 3 orders less energy and also has an N-shape CVC).

Further experimental and theoretical research has shown that autosynchronization is realized at the frequency of recombination resonance ω_{rc} (and its multiples)

$$\omega_{rc} = \frac{\omega_p + \omega_a}{2}.$$

In a close range telemetry communication channel the "active" immobile element works at antiresonance frequency, and the "passive" element (LC-circuit with a sensor on the moving object) works at resonance frequency [Sambyrski A.I., Novik V.K. Contactless measurements of parameters of rotating objects. Moscow, Mashinostroyenie, 1976].

For a better phase matching of two circuits it is advisable to use *impulse* electric excitation, with each impulse lasting $\Delta\tau \cong T/8$ (T is the oscillation period). The beginning of each impulse and its direction should coincide with the direction of the half-wave of the resonance circuit eigenfrequency oscillations (shock-and-vibration resonance). This impulse mode is used in mobile radiophones.

Our numerical estimates [Bogomol'nyi V.M. Theory of nondestructive testing in the electrical degradation of semiconductor devices. *Measurement Techniques*. 2001. Vol. 44. No. 5, p.513-516], as well as experimental data, have shown that the working mode for submicron field transistors is *the initial reversible stage* of electric degradation. The working frequency of submicron transistors has an intermediate value between the eigenfrequencies of *the electron and ion subsystems* in GaAs crystals.

The physical model of autosynchronization, formulated here, may be used for developing methods of non-destructive testing and eliminating technological defects in solid-state electronic devices [Bogomol'nyi V.M. Theory of nondestructive testing in the electrical degradation of semiconductor devices. *Measurement Techniques*. 2001. Vol. 44. No. 5, p.513-516].

Autosynchronization in solid-state electronic devices is realized at electric instability and negative differential resistance (NDR), when random fluctuations (especially those connected with structural defects) increase. The approximation of S- and N-shape CVC with cubic polynomials brings us to a generalized system of the van der Pol equations. If CVC can be Nor S-shape, the electrophysical processes in the nonlinear dynamic system are described by the Lienar's system of differential equations. If the reactance of the electric circuit is a function of voltage, we use the Abel's system of second kind. These mathematical models give a quantitative analysis of autosynchronization.

Studying fluctuations in autooscillation measurement systems has practical value for developing *time standards*. L.I.Mandelstamm was the first to study fluctuation in autosynchronization by observing the behaviour of nonlinear dynamic systems with random disturbances. Because autooscillation may include an arbitrary value of the initial phase, random "pushes" may "toss" the oscillation phase all over the place. The technique proposed here for autosynchronizing the oscillation phase can be used in the design of autogenerators and piezoelectric transducers, stable to the changing environment.

Synchronizing the performance of microcircuits used in computing is considered in: Kenniment D.I., Bystrov A., Yakovlev A.V. Synchronization circuit performance. *IEEE J. of Solid State Circuits*. 2002. V. 37. № 2, p. 202-209; Weste N.H.E., Eshraghiank K. Principles of CMOS VLSI design: A System Perspective. Second ed. Reading MA: Addison-Wesley. 1992, p. 326.

3.6. Mechanoelectroacoustic amplifier of sensor signals

A universal generator and amplifier of electric signals from sensors is constructed like an electromechanical transducer: *metal needle – ion crystal* (Si, Ge, GaAs), in which the current *self-oscillates regardless of the conductivity type and CVC of the crystal* (Fig. 18) [Melikyan E.G. Electroacoustic interaction on the metal-semiconductor contact. *J. Communs.Technology*. 1967. Vol.12, No.7 (translated from Russian)].

In this electromechanical amplifier, electric oscillations in the measuring circuit and *longitudinal mechanical oscillations* in the needle occur *simultaneously* and at *the same frequency* (from 0.5 to 600 kHz).

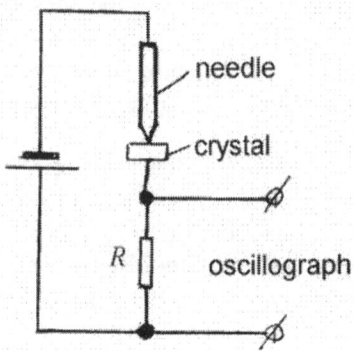

Fig. 28.

The microcontact of the needle point with the piezocrystal increases the sensor's sensitivity by *two orders*. As against tunnel and Gunn diodes, the static CVC of dielectric "needle-crystal" diodes *has no zone of negative "resistance"* (NR), i.e. has an *electroacoustical resonance character*. Polar semiconductors and DE play the role of the active element. At the micro-contact of the metal needle and DE occurs electron injection, which gives a thin-film DE semiconductor properties.

The PZ semiconductor crystal has an anomalously high tensosensibility. If *for any reason* the current through the diode increases, Joule heating raises the temperature of the needle point, thus elongating the needle and increasing its pressing force to the crystal. Therefore, the high tensosensi-bility of the diode *increases the initial current,* which creates *positive electroacoustic feedback* at the metal needle-semiconductor microcontact.

Amplifier oscillations were registered experimentally, with a quality factor of the electromechanical system $Q_{mech} = 15$. In long thin needles, the frequency of oscillations depends on needle length and ranges from 10 to 600 kHz. In short needles, longitudinal oscillations are excited at 500 Hz – 50 kHz. The shape of the oscillations is nearly sinusoid. Average cur-rent values are 22 mA (for silicon) and 80 mA (for germanium).

At relatively low supply voltage (when oscillations do not self-excite) one can obtain resonance amplifiers of *electric* and *"acoustic"* oscillations with an amplifying factor of 50. Such mechanoelectroacoustic amplifiers can be used in *acoustoelectronic generators,* supersensitive *vibration and noise sensors* (microphones) and viscosimeters (low-frequency range).

3.7. Mathematic model of porous MDM structures

The "electron nose", a unique innovation by Japanese engineers, contains a microprocessor and identifies the chemical composition of about 48 various substances in a gaseous medium. Porous dielectric structures with a highly developed surface ~100 m^2/cm^3 are used to make such chemical sensors more sensitive.

At the initial stage of the electrothermal breakdown the material is covered with micropores [71]. This process is accompanied by current oscillations in the MDM structure circuit and by electromagnetic radiation from DE into the environment. The information may be measured by standard radiotechnical devices and used for the non-destructive testing of technological defects (created deliberately to increase sensibility).

Let us determine the eigenfrequency of electromagnetic oscillations in a spheric cavity for a TM-wave (in which the radial component of the electric field strength is zero) from a wave equation [79, 108]

$$\Delta\vec{H} + k^2\vec{H} = 0, (1)$$

where $k = \dfrac{\omega}{c}$, k is the wave number, c is the velocity of electromagnetic waves in vacuum (speed of light), $\vec{H}$ is the vector of the magnetic field strength.

Exchanging Laplace operators Δ and *rot* gives

$$\vec{H} = \mathrm{rot}\,\vec{A}, \qquad (2)$$

where $\vec{A}$ is the vector potential, the equation of which in spheric coordinates has the form

$$\frac{1}{r}\frac{\partial}{\partial r^2}\left(r\vec{A}\right) + k^2\vec{A} = 0. \quad (3)$$

The solution of equation (3), finite at the origin of coordinates ($r = 0$), has the form

$$\vec{A} = \frac{\sin kr}{r}\vec{b}\exp(j\omega t), \quad (4)$$

where $\vec{b}$ is the constant vector, ω is the cyclic frequency.

The magnetic field strength is

$$\vec{H} = \left[\Delta\frac{\sin kr}{r}\vec{b}\right]\exp(-j\omega t). \quad (5)$$

The electric field strength $\vec{E}$ is determined by the Maxwell's equation

$$\text{rot}\vec{H} = -j\frac{\omega}{c}\vec{E}. \quad (6)$$

On the surface on the spheric cavity with an "a" radius electric resistance is zero; therefore, the tangential component $\vec{E}$ of the electric field $\vec{E}$ also becomes zero.

On the spheric surface $r = a$ the relation is realized

$$\left(\text{rot}\vec{H}\right)_{\theta} = \frac{1}{r}\frac{\partial}{\partial r}\left(rH_{\psi}\right) = 0, \quad (7)$$

where θ, Ψ are the angle coordinates in the meridian and azimuth directions. Because $\Delta\dfrac{\sin kr}{r} = k\dfrac{\cos kr}{r}\dfrac{\sin kr}{r^2}$,

$$r\vec{H}_{\theta} = k\cos kr - \frac{\sin kr}{r}. \quad (8)$$

In view of (6) – (8), the edge condition at $r = a$ for the tangential component of the electric field has the form

$$\frac{d}{dr}\left(k\cos kr - \frac{\sin kr}{r}\right)\bigg|_{r=a} = -k^2\sin ka + \frac{\sin ka}{a^2} - k\frac{\cos ka}{a} = 0. \quad (9)$$

Formula (9) gives us a transcendental equation

$$\text{ctg } ka = \frac{1}{ka} - ka. \quad (10)$$

The least root of equation (10) equals $ka = 2.74$, and the equality $k = \dfrac{\omega}{c}$ gives us an assessment of the eigenfrequency of the electromagnetic wave on the spheric surface

$$\omega_{min} = 2.74\,c/a. \quad (11)$$

For a GaAs crystal at $a = 50$ μm the eigenfrequency of oscillations calculated from formula (11) equals $f = 2.6\cdot 10^{12}$ Hz, which corresponds to experimental data on a $n - $GaAs diode.

Experiments show that at the electric breakdown stage in DE the velocity of hot electrons reaches $\sim 10^7$ cm/s. For gallium arsenide-based submicron field transistors in the working mode, the maximum velocity of electrons (the "ballistic" theory) is $\upsilon_{max} \approx 10^7 cm/s$. It follows that the working mode of these transistors entails a reversible electric breakdown and the formation of micropores.

Let us determine the critical size of pores a_{cr}, after which their growth becomes irreversible, by the condition of equal eigenfrequencies of the oscillations in the resonating cavity and solid-state plasma in the DE volume. Plasma frequency is calculated by the formula

$$\omega_p = \left(\frac{ne^2}{\varepsilon_0 m} \right)^{\frac{1}{2}}, \quad (12)$$

where ε_0 is the dielectric permeability of vacuum, e is the electron charge, m is the effective electron mass, n is the concentration of electrons injected into DE, determined by the expression

$$n = \frac{U \varepsilon_{33}^T \varepsilon_0}{2\pi h^2 e}, \quad (13)$$

where U is the electric voltage on the electrodes of the MDM structure, h is the DE thickness, ε_{33}^T is the relative dielectric permeability. The derivation of formula (12) is given in the Appendix.

Equating the left- and right-hand parts of equalities (11) and (12) in view of (13), we obtain

$$a_{cr} = 2{,}74 \, \mathrm{ch} \left(\frac{2\pi \, m}{\varepsilon_{33}^T U} \right)^{\frac{1}{2}}.$$

"Theory of lasing modes in microcavities and recent experimental data were presented. Microdisk lasers' supporting whispering gallery modes provide of large modification of the optical properties of the semiconductor, diode laser, whose wavelength is chosen for most efficient optical pumping of the microdisk"[89].

3.8. Calculation of the critical electric field strength in DE diodes

Measuring the curvature radius of the micropeak tip on the surface of the metal electrode is essential for estimating electrophysical characteristics of solid-state electronic devices. A similar task arises when we want to determine the concentration of electric and mechanical stresses in the electric insulation of electrotechnical devices, at the interfaces of materials in heterogeneous MDM structures, and near microcavities (pores) and rigid inclusions in DE.

Experimental study shows that one can *increase sensors' sensitivity* by *creating structural defects* in relatively "pure" ion crystals. The increase is explained by high concentration of electric and mechanical fields near structural defects (pores and cracks). Therefore, one should quantitatively assess the geometry of natural roughness at the metal-dielectric contact, which determines the concentration of electric, temperature and mechanical fields.

Under an electric field, microtips on the cathode surface inject electrons into DE. The interaction of the stationary flow of injected electrons with structural non-homogeneities, when solid-state plasma has low overheating instability, makes stationary oscillations of *the electromagnetic field self-excite* at the *resonance frequency of the defect*. Under an external electric field, defects may be brought into an electrically active state and become the sources of electromagnetic field fluctuations. Thermo- and autoelectron emission from the cathode and the instability of the electron subsystem lead to *an irreversible increase in fluctuations*, current self-oscillations and electromagnetic radiation from DE into the environment (at a wide frequency range from several kHz to hundreds of GHz, depending on the type and size of the defect). Some experiments registered elastic mechanical (acoustic) oscillations simultaneously with electric oscillations. Registering electromagnetic field oscillations may be useful for the non-destructive testing of solid-state electronic devices. Structural defects, especially those located *near the surface*, determine the mechanical, electrophysical and chemical properties of materials: electro- and heat conductivity, adhesion, adsorption, chemical reactivity (oxidation in air) [36].

The functional properties of thin-film heterogeneous microelectronic and electrotechnical devices (insulating layers) largely depend on the profile of natural roughness in metal electrodes, which under relatively weak electric fields causes electron injection into DE. As a result, DE contacting with metal in thin films acquires semiconductor properties. At the same time, in DE structures *low-temperature solid-state plasma* is formed. It is a system of injected electrons and ions (or atoms) of the crystal lattice (traps of free charge carriers on valent electron atom shells and in the energetic "prohibited zone" of DE).

Under an external electric field the electric field strength concentrates near micropeaks on the electrode surface, and in certain cases (depending on the size and shape of the tip) may exceed the average electric field strength by 2 orders. The concentration coefficients of mechanical stresses near holes, cracks, or rigid inclusions in a flat plate are determined by the Inglis's formula

$$K_t = \left(1 + 2\sqrt{\frac{L}{R}}\right), \qquad (1)$$

where L is the half-length of the hole or crack, R is the curvature radius of the hole at the point where the coefficient is determined.

The length of the crack can be measured precisely enough for practical purposes, whereas the curvature radius of the crack (the size of which may be comparable to the crystal lattice constant) presents certain technical difficulties.

Some experimental studies have looked into the concentration of the electric field strength near micropeaks on the cathode surface in MDM structures [Bogomol'nyi V.M. Theory of nondestructive testing in the electrical degradation of semiconductor devices. *Measurement Techniques*. 2001. Vol. 44. No. 5, p.513-516]. Using these experiments to calculate the maximum electric field strength in DE contacting with a micropeak on the electrode surface E_Z^{max}, one should employ a formula similar to (1),

$$E_z^{max} = \rangle E_z \langle \left(1 + 2\sqrt{\frac{H}{\rho_b}} \right), \qquad (2)$$

where H is the height of the peak on the metal surface, ρ_b is the curvature radius of the spheric surface of its tip, $\langle \mathbf{E_z} \rangle$ is the averaged (integral by DE thickness) electric field strength.

The value of the electric field strength $\langle \mathbf{E_z} \rangle$ in the right-hand part of formula (2) can be expressed by the critical electric field strength E_z^{cr}, which corresponds to the first cusp of the S-shape part of the static CVC. Near the cusp, under overheating instability in MDM structures, "current filaments" (10-30 μm in diameter) form and the current in the measuring circuit with a DE diode begins to oscillate.

Near the first cusp of the S-shape CVC the *temperature* of the sample *sharply increases*; subsequently, two or three current values correspond to one electric field strength. The MDM structure becomes unstable to overheating and the current oscillates. The critical field strength E_z^{cr} corresponds to the beginning of the irreversible stage of electrothermal destruction in MDM structures, accompanied by electromagnetic field oscillations.

Auto-wave processes caused by structural defects may be due to the equality of the *relaxation time of electron energy* interacting with crystal lattice τ_e (which characterizes electron energy dissipation at acoustic and optical phonons) and the time of dielectric relaxation τ_M.

The relaxation time of electron energy is

$$\tau_e = \frac{3}{2} \frac{n_0}{\sigma} \frac{\theta_0}{E_z^2}, \qquad (3)$$

where n_0 is the concentration of electrons injected into DE, θ_0 is the effective electron temperature in energy units, σ is the static DE conductivity under thermoelectron emission from the cathode.

The Maxwell's relaxation time τ_M (which generally characterizes changes in the electron, ion and structural polarization of DE) is calculated from the formula (see Appendix 1)

$$\tau_M = \frac{\varepsilon_z}{4\pi\sigma}, \quad (4)$$

where ε_z is the dielectric permeability of the crystal.

Equating the left- and right-hand parts of expressions (3) and (4), we obtain the bottom boundary assessment of the critical electric field strength E_z^{cr}, which brings about an irreversible electric breakdown

$$E_z^{cr} = \left(\frac{6\pi n_0 \theta_0}{\varepsilon_z}\right)^{1/2}. \quad (5)$$

The average volume density of electrons injected into DE is calculated by the formula [1]

$$n_0 = U\varepsilon_z / 2\pi h^2, \quad (6)$$

where U is the electric voltage on the electrodes of the MDM structure, e is the electron charge, h is the DE thickness.

Substituting (6) into equality (5), we obtain:

$$E_z^{cr} = \frac{3\theta_0}{he}. \quad (7)$$

It follows from (2) that in reality E_z^{cr} in the left-hand part of (7) is an average integral characteristic and should be multiplied by the expression in brackets to the right of (2). One should also take into account the concentration of the electric field strength near structural non-homogeneities. Thus, it follows from (7) that

$$E_z = \frac{3\theta_0}{he}\left(1 + 2\sqrt{\frac{H}{\rho_v}}\right)^{+1}. \quad (8)$$

Using equality $E_z = U/h$, we obtain from (8)

$$U = \frac{3\theta_0}{e}\left(1 + 2\sqrt{\frac{H}{\rho_v}}\right)^{+1}. \quad (9)$$

Under given electric voltage on the MDM structure electrodes U, by measuring the temperature on its surface, we can calculate the ratio H/ρ_v when the current begins to oscillate in the electric circuit. If it is possible to calculate experimentally the average statistical height of micropeaks H, formula (9) can give us the curvature radius ρ_v.

To assess the shape and size of the structure is necessary not only for developing methods for the non-destructive testing of materials and constructions, but can also help to eliminate technological microdefects, inevitable in the industrial manufacture of electronic and electrotechnical devices. In case of overheating instability, secondary emission electrons, positive ions and electrons injected into DE form low temperature solid-state plasma. When the concentration of negative and positive charges is the same, one can use hydrodynamic analogy. Formulas (5), (9) are obtained in the approximations of the hydrodynamic model of solid-state plasma [Bogomol'nyi V.M. Theory of nondestructive testing in the electrical degradation of semiconductor devices. *Measurement Techniques*. 2001. Vol. 44. No. 5, p.513-516].

Apart from offering *CVC analysis,* we also demonstrate that the geometry of structural defects can be estimated by *temperature measurements.* These results can also be used in the manufacture of submicron permeable base field transistors (for oscillation generators and amplifiers of weak signals in sensors), since these need 3 or more orders of magnitude less energy than ordinary field transistors.

The proposed technique for estimating the micropeak curvature radius may help to develop non-destructive testing for lining sapphire plates with artificial diamond films. This technique may also be useful in the fabrication of injecting diamond cathodes for scanning tunnel microscopes (STM) [Olson B.W., Chen K. Microstructure defect detection using thermal response. *Proc. SPIE.* 2002. V. 4755, p. 714-725].

3.9. Electrophysical study of structure, mechanical stresses and strains

Integral circuits in microprocessors use hyperthin films (HTF) with high *plasticity.* The plasticity is the higher, the easier dislocations move under mechanical load. Most dislocations move through HTF from one external surface to the other. Therefore, they are very short and can be easily displaced under small loads. The ultimate stress σ_{fr} (fracturing stress) and relative elongation before rupture increase along with plasticity. For example, for 100 μm gold wire $\sigma_{fr} = 150$ MPa, for 200Å –thick membrane (0.02 μm) $\sigma_{fr} = 530$ MPa. The concentration of disk-shaped cracks at plastic strain in aluminum is approximately n = 10^9 cm^{-3}. *Elastic and plas-*

tic strains occur stepwise. At brittle destruction highway *cracks spread at the speed of sound*. The velocity vector "pulsates" both in size and direction (crack branching).

Destruction is a kinetic process of *failure accumulation*, which occurs throughout the whole loading (mechanical, electrical and temperature) period and is not a critical event of losing mechanical stability at the moment of destruction. When the top of the highway crack moves stepwise through the "forest" of dislocations (for GaAs crystal the velocity $v = 2.3 \cdot 10^3$ m/s), the sign of mechanical stresses changes periodically and quickly, which raises *waves of "unloading"* (release of elastic energy accumulated at the half-wave of elastic compression). These elastic waves come out of the crack top and its *newly formed* surfaces and are registered by acoustic emission techniques. They are used in industry for the non-destructive testing of materials and constructions [Seryoznov A.N. et al. Negatronics. Novosibirsk. Nauka, 1995].

Mechanical destruction causes an outburst of high-energy *electrons* and *interatomic* radiation such as X-ray and gamma-ray photons. For example, mica lamination produces X-ray radiation, and sugar cracking gives electron outbursts with 30-40 keV .

Microcracks appear *well before destruction*. Usually they originate on the surface and greatly contribute to electric resistance at high frequencies (when the current moves near the surface). *Cracks increase electric resistance*.

The durability of metal parts in machine building is generally predicted by the eddy currents technique (skin effect) [108].

One of the simplest and at the same time precise non-destructive methods for studying degradation and predicting toughness of materials is measuring the *electrical resistance of metals, semiconductors and dielectrics* [Vladimirov V.I., Lupashku R.G. Crack study by electroresistance technique. *Strength of Materials*. 1973. No.4, p. 70-74]. Long range testing of DE is performed by measuring capacity (dielcometry).

The scanning tunnel microscope (STM) is the most precise device for studying the molecular structure of materials. Its principle is the following: a needle is attached to a console-mounted multi-layer piezoceramic plate and emits electrons. The density of the emission current displays the interatomic structure of the material.

The needle is discretely moved along the surface of the sample by piezoelectric motors. The spacing is comparable to the crystal lattice constant.

Linings in memory devices are based on a transparent monocrystal of gadolinium gallium garnet (GGG). The main requirement is the absence of residual technological mechanical stresses.

The plates for linings are fabricated by cutting an ingot. Ingot pieces with residual stresses have to be cut out and recycled. The waste of this expensive material amounts to about 30%.

Non-stressed parts in optically transparent materials are usually located by optical polarization techniques. This method cannot be used for GGG because of its relatively high photoelastic anisotropy (k = 0.38). On the other hand, its elastic anisotropy k = 0.95, and we can use *acoustic tensometry* [24, 29, 35, 38; Seryoznov A.N. et al. Negatronics. Novosibirsk. Nauka, 1995].

Acoustic techniques for the non-destructive testing of mechanical stresses measure the velocity of US waves and give no reliable results if the sample has a complex shape (with a bent outer surface). To find out how mechanical stresses are distributed in a GGG monocrystal both with flat and cylindrical surfaces, researchers use the *echo impulse technique*. This method allows one to calculate the value and distribution of main normal stresses by the number of the echo impulse in a sequence of reflections.

Construction strength is usually measured with tensosensors. There are microwave (MW) methods for measuring non-electric values, in which transducers are electromagnetic elements with distributed parameters [Seryoznov A.N. et al. Negatronics. Novosibirsk. Nauka, 1995]. The measuring system registers the changes that happen in the material when MW interact with the electromagnetic field. MW primary transducers analyse the dependence of the reactive component of the IC input impedance on the physical value measured $Z = R(t) + jX(t)$, with $R(t) = 0$.

This is the difference between distributed measuring electromagnetic transducers and resistive transducers, in which information is received through the *active component* of the electric *impedance* $Z - R(t)$ (in numerous discrete points of the sample). At present preference is given to resonance transducers with pronounced frequency-selective characteristics. Primary distributed transducers offer radically new possibilities because their size is comparable to that of the sample and they can be electronically scanned lengthwise. For example, one can measure temperature fields or the distribution of local deformations both on the surface of the object and in the volume of the material. Distributed transducers use *only one connection line,* and the frequency of electronic scanning can reach hundreds of kHz. This is equivalent to parallel use of several hundreds of discrete transducers based on various physical principles of transforming physical values into electric signals. One can supplement them with any radiolocating devices to measure the distance to the point that reflects an electromagnetic wave. Parts of long electric lines excited with MW currents are used as primary strain transducers and identifiers of local failures of machine elements. They determine the moment of destruction, the location and size of defects in the material, e.g. crack coordinates.

Modern information measuring systems (IMS) measure real-time changes of the *surface profile*. The physical essence of transforming *the geometric image of the profile* into a measuring *signal* is the quantitative analysis of distribution and dissipation of the impulse electromagnetic field in the electric line formed between the surface under analysis and a distributed sensor.

The mathematical model for transforming the profile into an electric signal is given in [Seryoznov A.N. et al. Negatronics. Novosibirsk. Nauka, 1995]. It is based on the Maxwell's electrodynamic equations and the theory of non-homogeneous lines, which is widely used in radio technology.

3.10. New materials and technologies in microelectronics

A modification of the scanning tunnel microscope (STM) can be used for implanting ions into crystals and for monomolecular discrete coating.

Tip nanolithography is a new direction in the technology of manufacturing heterogeneous structures, i.e. microcircuit elements [Johnson K.H., Pepper S.V. Molecular orbite model for metal-sapphire interfacial strength. *J. Appl. Phys.* 1982, v. 53, No. 1, p. 6634-6637]. STM was used for adsorption of metal atoms of Ti, Cr, Cu and Al onto monocrystal films Al_2O_3 (sapphire), MgO.

The effect of luminescence on *autoelectron emission* (AEE) in carbon materials can be used for light-emitting diodes, computer displays (scintillators), dosimeters of ionizing radiation, sensors of the electromagnetic field strength or optical and laser radiation.

AEE is an "outburst" of electrons under an external electric field. Its distinguishing feature is *the absence of energy consumption* [Obraztsov A.N., Volkov A.P., Pavlovsky N.Y. Mechanism of cold electron emission from carbon materials. *Letters to Sov. Physics – Technical Physics.* 1998. Vol. 68. No. 1, p. 56-60].

The reduction of the electron work function may be caused by the following reasons:

- decrease of the potential barrier on the emitter surface due to the *quantum mechanical tunnel effect,*
- use of *diamonds* as the source of free electrons owing to their negative electron affinity (NEA) (found in diamond monocrystals and polycrystal films obtained by gas phase chemical deposition),

- reduction of the threshold electric field strength, necessary for low voltage AEE. It is used in autocathodes consisting of carbon (carbon C_{60} modification) nanotubes and *graphitelike* nanostructures, as well as in plate-like crystallites on film surfaces due to the concentration of the electric field on carbon microtips.

This simple *technology of film electron emitters* (the gas phase epitaxy technique) is also of great practical interest. Relatively long (up to 100 μm) nanotubes on the emitter surface with a diameter from several angstrom to dozens of nanometers (10^{-3} micrometers) yield AEE at small potential difference, because the coefficient of the electric field strength concentration K_t is proportionate to the square root of the ratio H to ρ_t (where H and ρ_t are the height and curvature radius of the nanotube top).

Traditional methods for studying the microstructure of materials (selective etching, optical and electron microscopy, X-ray techniques) have no time resolution and do not show the evolution of the structure. The dynamics of mesoscopic defects in DE can be studied by *electromagnetic emission*. This technique helps to identify the three-dimensional structure of electrically active structural defects, slipbands and microcracks, as well as to obtain amplitude-frequency characteristics of growing structural defects.

Microelectronic components (from micron fractions to 5-10 μm), inexpensive light-emitting panels and displays can be manufactured from organic polymers in the liquid state, which after solidifying acquire special electrophysical, in particular electroluminescent, properties. Thin polymer films are sprayed on transparent plates with piezoceramic jet printers (used for computer information output) [1].

Direct write technology (DWT) greatly simplifies the construction of microelectronic devices [Szczech J.B. et al. Fineline conductor manufacturing using drop-on-demand PZT printing technology. *IEEE Trans. on Electronics and Packaging Manufacturing.* 2002. V. 25, No. 1, p. 26-33]. DWT is used in the matrix microbolometer (10×10 elements matrix) with separate 60 × 60 μm sensitive elements. A bolometer contains semiconductor or pyroelectric elements, which are sprayed with a jet printer onto a 125 μm elastic polyimide film. Bolometers are inexpensive and have the following characteristics: temperature resistance coefficient– 3.07%/K at room temperature, sensibility 3×10^7 cm(Hz)$^{1/2}$ /W at $f = 1.13$ Hz and current ~1 microampere [Yaradanakul A. et al. Uncooled infrared microbolometers on a flexible substrate. *IEEE Trans. on Electron Devices.* 2002. V. 49, No. 5, p. 930-933].

Ordered nanostructure clusters with a characteristic de Broglie wave length (in the volume and on the surface) were called quantum wires and quantum dots. They are used in the newest micro- and opto-electronic devices.

Electromechanical systems for computer graphic information output and book printing include *ink-jet printers*, which print colour photographs. Two-dimensional images are created by displacing charged paint drops in an electromagnetic field. Physical models of the formation of *charged drops* (and the destruction of ion crystals) were proposed by Y.I.Frenkel [Mechanism of solid and liquid electrization at dispersion. *J. Exper. Theor. Physics.* 1948. Vol. 18. No. 9, p. 799-806]. Jet printers are used in manufacturing thin-film solid-state heterogeneous structures (microcircuits).

Biological microobjects (cells and their nuclei, micromolecules), small quantities of substances for cellular microinjections are studied with micromechanical instruments manufactured by the semiconductor technique with the help of micropositioning piezoelectric devices (piezoceramic actuators).

The researchers from the Institute of semiconductor physics of the Russian Academy of Sciences have designed and built chip microlaboratories, silicon microneedles for injecting miniscule amounts of liquids into nervous tissues (100 Å to 1 μm in diameter), microscalpels (which extract individual cells).

The manufacture of monocrystal semiconductor nanotubes is based on the self-rolling of thin heterogeneous films. The films can be exfoliated from the substrate under internal mechanical stresses. The diameter of nanotubes depends on film thickness and mechanical stresses. The size of the cross section is maintained with a high precision at molecular-beam epitaxy ranging from 30 Å to 100 μm. At present the minimal thickness of nanotubes is 5 Å.

The proposed technology was used to develop microelectrodes and microsyringes for transferring and measuring very small quantities of liquids and microparticles (up to 10^{-15} g). The monomolecular films were based on GaAs, InGaAs, AlAs crystals and heterostructures.

Integral microelectronics, measuring and control systems contain semiconductor integral microcircuits (SIMC). The substitution of SIMC for LIC and VLSI gives the best economic effect. However, structural imperfections of silicon crystals may limit their application.

Silicon is a brittle material with small tension and bending strength, but it is very hard. When ingots are cut, lapped and polished, the monocrystal is covered with *cracks and scratches*.

The manufacture of silicon plates includes multiple mechanical treatment, which causes residual stresses in surface layers.

Because the face (working) and "reverse" side of the monocrystal are treated non-symmetrically, residual stresses are distributed along the crystal thickness unevenly. This causes the plate to deform, bulging towards the rougher side.

Stresses in the near-surface treated layer σ_s and in the substrate volume σ_v are calculated by the formulas

$$\sigma_s = \frac{E_{elast}}{6(1-v)} \frac{h^2}{t} \left(\frac{1}{R_1} - \frac{1}{R_2} \right),$$

$$\sigma_v = \frac{E_{elast}}{1-v} \left(z - \frac{2}{3}h \right) \left(\frac{1}{R_1} - \frac{1}{R_2} \right),$$

where E_{elast}, v are the Young's modulus and Poisson's coefficient, h is the substrate thickness, t is the thickness of the damaged layer, z is the co-ordinate marked off the interface of the damaged layer and the remaining part of the substrate, R_1 and R_2 are the curvature radiuses of the surface before and after mechanical treatment.

Microcircuit elements have micron and submicron sizes and a high degree of integration ($10^6 - 10^7$ cm^{-2}). The defect sizes are comparable to the sizes of LIC elements.

Defects (from point defects to clusters and dislocations) cause lattice strains and, therefore, mechanical stresses. It is possible to decrease the concentration of structural defects in LSI and VLSI by *gettering* (Kang J.S., Schreder D.K. Gettering in Silicon. *J. Appl. Phys.* 1998. V. 65. No. 8, p. 2974-2985).

4. Negatronics

4.1. Devices with negative differential resistance

Our knowledge in all fields of science depends on *the characteristics of measuring technology*. Automated information measuring systems are based on programme-controlled microprocessors, oscillating LC-contours, generators, filters and piezoelectric transducers, i.e. sensors of various physical values and chemical composition [29, 46, 47, 62, 83, 102; Seryoznov A.N. et al. Negatronics. Novosibirsk. Nauka, 1995].

Very large scale integrated circuits (VLSI) in microprocessors and single-chip computers are manufactured as integrated *inductivities* and *capacities*. To achieve the maximum density of VLSI components and simplify their manufacture, one can use *negatrons* with negative differential resistance (NDR), i.e. *semiconductor analogs of inductivity* (SAI).

Planar technology in microelectronics (photo and X-ray lithography) allows one to synthesize integrated standard transistors and other active elements with NDR. Unlike manufactured NDR devices (tunnel and Gunn diodes), *transistor inductivity analogs* have an electric resistance that can quite easily be changed in a wide range (it is possible to obtain *negative, zero and positive electric "resistance"*).

Negatrons are characterized by *current and voltage positive feedback* (PF). This determines the capacity (or inductive) nature of reactive resistance, as well as the CVC shape: S-, N- or λ-type. If a device has a *voltage* PF, its reactivity is *capacity*, and its CVC is N-shape. A device with a *current* PF *has inductive reactivity* and an S-shape CVC.

MDM structures based on *polar DE and FE* offer the simplest way of controlling voltage and turn-on/off current and separating input and output LIC circuits, which considerably *expands their information capacities*.

The *MDM* structures that work at the *initial reversible* stage of electric breakdown (without mechanical destruction of DE) realize NDR in the simplest way. Functional characteristics of *homogeneous DE structures* in the modes of *thermal and autoelectron emission* greatly vary depending on the controlling electric field [Bogomol'nyi V.M. Calculation of energy dissipation and heating temperature in piezoelectric MDM structures. *Measurement Techniques*. 1998. No. 12; Bogomol'nyi V.M. Theory of nondestructive testing in the electrical degradation of semiconductor devices. *Measurement Techniques*. 2001. Vol. 44. No. 5, p.513-516. www.wkap.nl/journals/mete].

At the reversible stage of electric breakdown MDM structures near micropeaks on the electrode surface are covered with channels of electron conductivity −"current filaments" (without any macromechanical destructions of the DE layer). Under certain conditions the flow of electrons injected into DE is compensated by the positive space charge (near current filaments) and forms *quasineutral solid-state plasma*. This plasma contains a *spiral electron* flow (tornado) directed at the vector of the electric field strength. The integral characteristic of the flow is solenoidlike (see Appendix 2). Thus, MDM structure with an S-shape CVC forms a negatron with inductivity (current filaments) and capacity (the volume of a polar DE). It should be noted that the equivalent circuit of the piezoelement contains inductivity and capacity, i.e. forms an LC-circuit [2].

Field-controlled switches and memory elements in microprocessors are manufactured from sandwich structures such as MDM and MDS (nonlinear analogs of standard dipole and field transistors). These devices can be built by *introducing special defects* into MDM structures (admixtures, dislocations, micropores, regular controlled roughness of the electrode surface). Porous polycrystal silicon films with fragments of silicon monocrystals (or SiO_2) are used in active elements of logical and analog IC for amplifying and processing signals, in resistors, heterogeneous solid-state structures with inductive resistance (two-base diodes) and negative capacity [Seryoznov A.N. et al. Negatronics. Novosibirsk. Nauka, 1995].

Negatrons are used as *measuring transducers* in MW systems for assessing the durability of aviation constructions (to find crack coordinates and sizes, to test the profile of surface roughness), parts of friction units in air-tight refrigerator compressors and car engines.

Negatrons can be used in electronic-mechanical *time devices* and precision frequency measuring devices, in automatic control systems in refrigerators, washing machines, tape recorders, as well as in board and stationary *diagnostics systems* in automobiles.

Tunnel diodes have the following disadvantages:
- low level of working voltage,
- low output power,
- incompatibility with IC manufacture technology,

- *unstable* characteristics.

Generator and key circuits in fast analog-digital transducers (where high metrological characteristics are unnecessary) contain *avalanche transistors* (AT). They have high performance ($f \geq 400$ kHz) and an NDR of dozens or hundreds Ohm. *Avalanche phototransistors* work *2 – 3 orders faster* than standard phototransistors and phototiristors and need dozens or hundreds V to enter the conducting state. Their parameters are varied; their activation requires a complex system of current (or voltage) regulation. These disadvantages *limit AT applications*.

Multivibrators, relaxation generators, frequency dividers, protection devices, time relays are based on *unijunction NDR transistors* (kOhm). These transistors have *drawbacks similar* to AT. *NDR devices eliminate these disadvantages and are used in voltage stabilizers and overload protection systems*. A comparative analysis of *manufactured NDR devices* shows that dinistors, tiristors, unijunction and avalanche transistors do work in the amplifier, generator and key modes, but are mostly applied in *impulse technology*. This is explained by the following disadvantages:

- low temperature stability,
- instability and high sensibility of the transformation coefficient to a change in NDR,
- *high* supply voltage and consequently problematic connection with analog microcircuits and sensors.

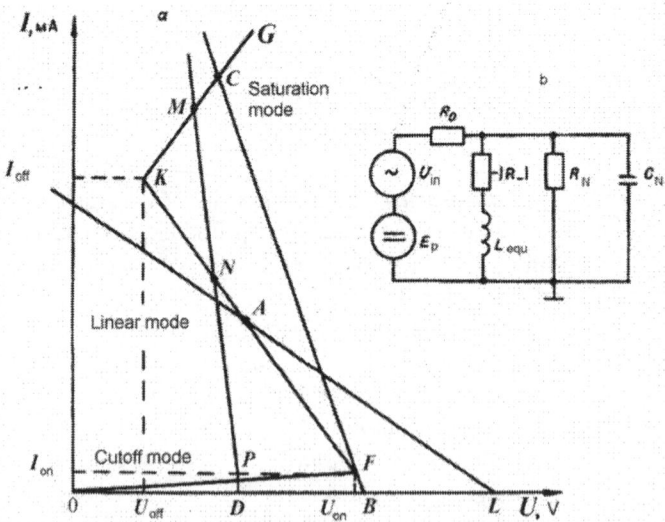

Fig. 19. Choosing the working point on an S-shape CVC (a) and external circuit connection of a two-terminal element with an S-shape CVC (b). [Seryoznov A.N. et al. Negatronics. Novosibirsk. Nauka, 1995].

Fig. 19 features a circuit with an NDR device $|R_-|$ and a sequence resistor R_0, which gives the working mode and location of the "working point". In the OF part of the CVC (Fig. 19) the transistor works in the cutoff mode; FK is the linear mode, KG is the saturation mode. If *the load line* MD crosses the S-shape CVC in three points, the device plays the role of a *trigger*. If the load line AL crosses the CVC in *one point A*, the device functions as an *amplifier* or *generator*. If the line BC crosses the CVC in *two points*, the device works in the *threshold (key) mode* (i.e. one device performs various functions).

In the OF part of the CVC (Fig. 20), transistors are closed (the current is almost absent). *The opening voltage approaches breakdown voltage.* Fig. 20 shows a transistor equivalent of the p-n-p-n structure (thyristor). When transistor V2 is switched on, the current moves through its collector circuit, and the voltage on the resistance of the collector circuit drops. This opens transistor V1 (current positive feedback) (curve 1, Fig. 20).

Increasing supply voltage E_s increases the current through the resistive divider R_{k1}, R_{bb}, R_{k2} and decreases the voltage on the collector resistors R_{k1} and R_{k2}.

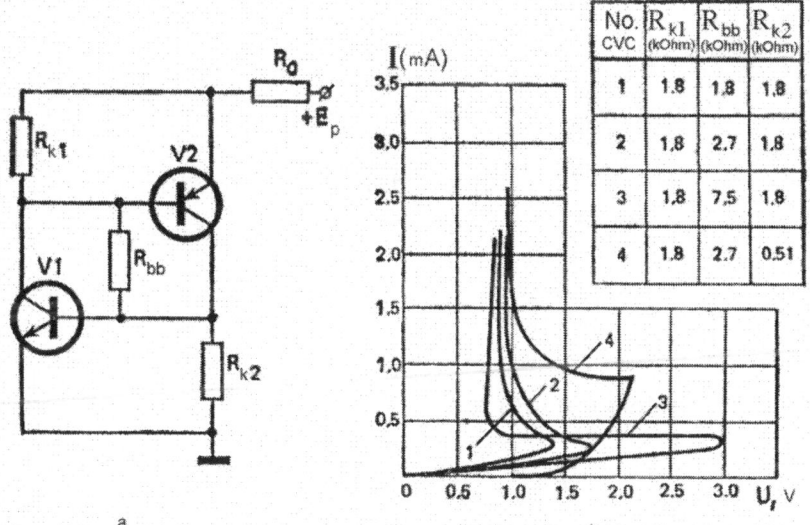

No. CVC	R_{k1} (kOhm)	R_{bb} (kOhm)	R_{k2} (kOhm)
1	1.8	1.8	1.8
2	1.8	2.7	1.8
3	1.8	7.5	1.8
4	1.8	2.7	0.51

a

b

Fig. 20. (a) is the layout of a transistor analog of the p-n-p-n-structure with a voltage negative feedback; (b) shows CVC devices [Seryoznov A.N. et al. Negatronics. Novosibirsk. Nauka, 1995].

When voltage on these resistors falls equally, the base-emitter junction U_{be1} (U_{be2}) of one of the transistors opens. Current PF immediately opens the second transistor (see Fig. 19, b). Since the transistors have a scattering of parameters, the first to open is the transistor with a smaller resistance on the base-emitter junction (U_{be}).

The resistance of the resistor R_{bb} is deliberately much larger than R_{K1} and R_{K2}. Hence, the power of the current through the resistive divider mainly depends on R_{bb}.

In the linear mode, the current in the common circuit flows mostly through transistor collector circuits. When the transistors are in the saturation mode, the resistances of the base-emitter junctions shunt the collector resistors. Currents in PF circuits decrease to zero and flow in the opposite direction. When the currents in PF circuits are zero, a negatron switches off.

Multistable NDR devices

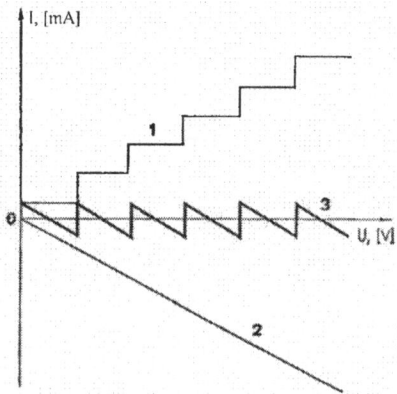

Fig. 21. The resulting CVC of a device with n NDR zones [Seryoznov A.N. et al. Negatronics. Novosibirsk. Nauka, 1995]

There are simple and reliable *computing devices* (operating as current and voltage stabilizers, generators of multiple frequencies, amplifiers with a variable amplifying coefficient) and *memory elements with high informational capacity,* which are negatrons with several stable NDR zones (this can reduce the number of elements in the circuit). It is possible to make n NDR zones on the CVC by connecting *n* nonlinear dipoles parallel to a device with one NDR zone. Nonlinear dipoles should consist of coupled circuits with a current and voltage stabilizers. In each couple the stabilizers are connected either in series or parallel.

When current and voltage stabilizers are *in series*, separate circuits are *in parallel*, and vice versa.

The joint CVC (kinked curve 3, Fig. 21 is a sum of the CVC of each circuit and the CVC of the NDR device) is curve 2.

4.2. SAI. Temperature drift compensation

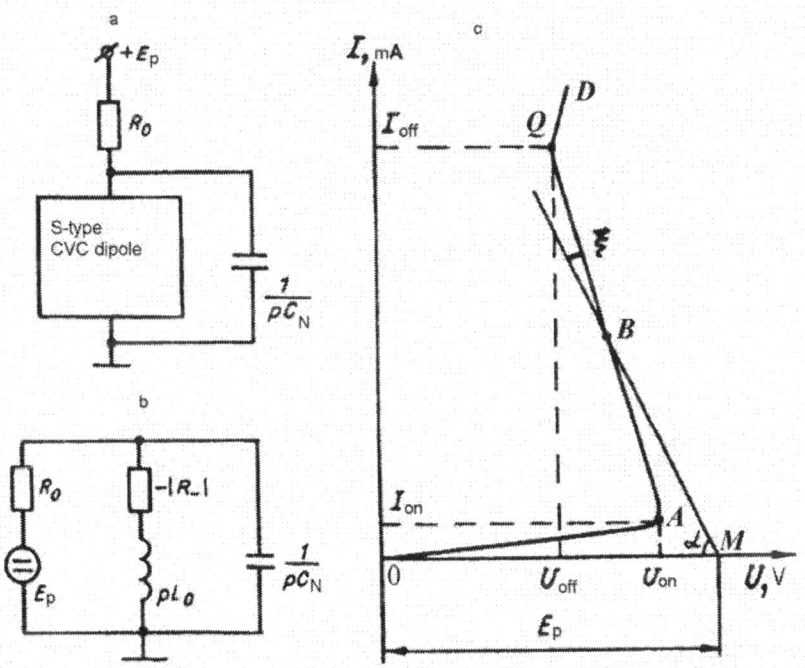

Fig. 22. Functional layout of SAI based on a dipole with an S-type CVC (a); SAI equivalent circuit (b) and the CVC showing the principles of SAI with a negative resistance (NR) element (c) [Seryoznov A.N. et al. Negatronics. Novosibirsk. Nauka, 1995]

SAI in a microcircuit perform the function of wire inductivity. However, they *do not "store" magnetic energy*. SAI are *not affected* by power *transformers'* fields or other sources of electromagnetic noise. The small size (and mass) of a SAI is achieved by *planar technology*. Transistor layers are connected with the surface of the semiconductor plate, which allows fabricating contacts and easily interconnecting them in one electric circuit. Thermocompensation is realized by a logistic processor with temperature, voltage and current sensors.

The functional and equivalent circuits of thermocompensation with negatrons are shown in Fig. 22 a, b. The inductivity of SAI (NR $|R_-|$) is connected with the resistance of the current-driving resistor R_0 and the load capacity C_L. In the working mode the loading line MV crosses the CVC of the resistor R_0 at the point B (Fig. 22 c). Thermocompensation is achieved by changing the ratio of the "depth" of *the current positive and voltage negative feedback*.

4.3. Physical model of SAI (equivalent circuits technique)

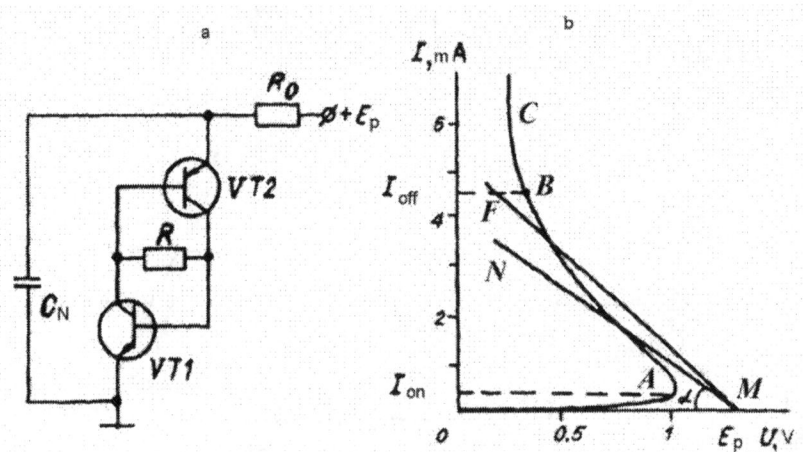

Fig. 23. Layout of SAI based on the simplest equivalent of p-n-p-n-structure (a) and the CVC explaining the changes of NR $|R_-|$ at varied resistance of the resistor R_0, which gives the working point (b) [Seryoznov A.N. et al. Negatronics. Novosibirsk. Nauka, 1995]

Semiconductor appliances and *devices with S-type CVC and inductive reactivity can be used as SAI*. The most promising among them are thyristors and dynistors with a *p-n-p-n*-structure. Semiconductor elements based on *p-n-p-n*-structures can be divided into two groups. One group uses complementary pairs of *p-n-p-* and *n-p-n-*transistors with an internal current PF, which is controlled by the external feedback. The devices of the other group contain at least two transistors with an external current PF. It should be noted that in the former the bases of the complementary transistors have floating potential bases and in the latter *the voltage on the transistor bases is fixed by an external energy source*.

The semiconductor devices of the first group are used for building SAI. They are connected in series with the resistor R_0 and the energy source E_s (Fig. 23). By varying the resistance of the resistor R_0, we choose the working point on the NR segment of the CVC (see Fig. 23, b). If the resistance of the resistor R_0 changes, it also changes the inclination of the load line $\alpha = \text{arctg} R_0$. Since an S-type CVC is nonlinear, moving the working point along AB varies the value of NR $|R_-|$, and, consequently, the inductivity $L = |R_-| t_3$ (t_3 is the delay time of current changes, caused by current PF). The resistance of the load resistor R_0 compensates NR $|R_-|$ and raises the quality of SAI (operating speed and "sensibility"). To work steadily, SAI needs the condition of inequality (stability by the constant current)

$$R_0 > |R_-|. \quad (1)$$

If the inequality sign changes ($R_0 < |R_-|$), the device generates sustained oscillations, and the load line MF crosses the working zone of CVC at one point.

The quality of SAI is determined by the formula

$$Q = \frac{\omega L}{R_0 - |R_-|}. \quad (2)$$

When $R_0 \rightarrow |R_-|$, the quality grows indefinitely, but functional stability abruptly decreases.

In real SAI circuits the quality $Q = 50 - 100$. The boundary evaluation of operating stability at constant current is given in the expression

$$L = C_L R_0 |R_-|, \quad (3)$$

where C_L is the load capacity.

The condition of stability at alternating current

$$L > C_L R_0 |R_-|. \quad (4)$$

It follows from (1)-(4) that NR $|R_-|$, the inductivity L, Q and the stability reserve are connected with the delay time t_3, which depends on the "depth" of current feedback.

These negatron circuits are used in industrial semiconductor transistors and resistors. The USA and Japan have developed dielectric GaAs-based diodes and transistors with nonlinear S-characteristics (and N-type CVC), which operate at *the initial reversible stage of electrothermal breakdown* (without mechanical destruction). In DE electronic devices CVC change due to *the injection of free charge carriers from the electrodes*. E.g., in a crystal GaAs submicron field transistor with a permeable base (PBST) electrons move "ballistically" (without "colliding" with the atoms of the crystal lattice). The mathematic model of a submicron field transistor is given in [Bogomol'nyi V.M. Theory of nondestructive testing in the electrical degradation of semiconductor devices. *Measurement Techniques*. 2001. Vol. 44. No. 5, p.513-516. www.wkap.nl/journals/mete] with formulas for calculating its parameters (the frequency and electric field strength for the self-synchronization of resonance oscillations).

Polycrystal polar DE and high ohmical semiconductors have certain advantages compared to monocrystals. Opto- and microelectronic devices are based on optically transparent electrostrictive PZLT ceramics with the quality $Q_{mech} = 50 - 100$, dielectric permeability $\varepsilon_{33}^T \cong 2000 \div 3000$, and large ohmical resistance. They also operate *one order faster* than their semiconductor counterparts.

5. Modern measurement and information processing techniques

In 1922 A.Einstein was awarded the Nobel Prize in Physics for the photoelectric law and his works in the field of theoretical physics.

As discovered by W.Hallwachs and H.H. Hertz in the 1880 s, incident light can eject electrons from metals. The *velocities* of these *photoelectrons* are *independent* of the intensity of the light but *increase with its frequency.*

In 1905, A.Einstein proposed a physical picture of the photoeffect, in which the light acts as quanta of energy hv, where h is Planck's constant and v is the frequency of the light (v = C/λ, λ is the wavelength), The proposed equation of kinetic energy contains no reference to the intensity of the light, but gives the *energy of the ejected electrons in terms of the frequency only.*

Studying the way electromagnetic waves move in crystals, A.Einstein found that their radiation frequency equals absorption frequency. This principle became the basis of lasers.

Hence the universal algorithm for healing all technological structural defects at the initial reversible stage of electric breakdown. At this stage the eigenfrequencies of the structural microdefects are measured (current oscillations, electromagnetic radiation) with subsequent healing by the external electromagnetic field.

Photodiodes, photoelectron multipliers, electron-optical transducers and transmitting TV tubes are some applications of the external photo effect, which was discovered by Herz in 1887. In 1905 A.Einstein explained its quantum-mechanical nature.

Covering atomically clean p-GaAs surface with cesium-oxygen creates the effect of negative electron affinity (NEA). The energy level of vacuum is located below the conductivity zone in the semiconductor.

Photoelectron emitters with NEA have unique characteristics: the quantum yield of the external photoeffect reaches 0.4 – 0.5; there may appear spin-polarized electrons. The electron beams have a narrow angle and energy distribution. NEA photoemitters are used in molecular and atomic physics [Alperovich V.L. et al. Domination of adatom-induced surface states on p-type GaAs-Cs-O at room temperature. *Phys.Rev.B.* 1994 Vol.50. No.8, p.5480-5483; Orlov D.A. et al. New tool for detailed study of NEA-photocathode parameters including spin-polarization. *Proc. Of 12th Int.Symp. on High-Energy Spin-Physics.* Amsterdam. The Netherlands. 1997, p.717-719; Alperovich V.I. et al. Evolution of surface electronic properties of GaAs photocathodes during degradation. Ibid. p.755-757; Turnbull A.A., Evans G.B. Photoemission from GaAs-Cs-O. *J.Phys.D: Appl.Phys.* 1968. Vol. 1. No.2, p.155-160].

The physical nature of molecular beam epitaxy, nanolithography for ultra-small structure fabrication, resonant-tunneling devices, hot electron transistors and ballistic electron transport and their quantum technologies are considered in [70] and [Bogomol'nyi V.M. Resonance calculation for electrothermal damage to metal-insulator structures. *Measurement Techniques*. 2000. Vol.43. No.6, p.538-543].

Conclusion

Real life is so much more complex than conventional physical models that the cognitive process is just a thorny way to the Truth. Nevertheless, the understanding of the boundaries of reasonable sufficiency, the method of analogy, experimental data, James Clerk Maxwell's electrodynamic equations, electroelasticity relations and the concept of the wave nature of the electron do allow a qualitative picture of physical phenomena in sold-state electronic devices.

This book considers classical examples of the self-synchronization of oscillations. We have studied the initial reversible stage of electric breakdown of DE in contact with metal, using the theory of nonlinear dynamic systems. This gives us an understanding of the resonance wave nature of processes in all diodes, transistors and integrated circuits, as well as in the keys and generators of mobile radiophones (when the principle of minimal energy is realized).

Solid-state physics is well aware of dimension effects, such as Debye screening radius, electron wave length and run time. This book determines the diameter of the spherical area of interaction of the electron with the atoms in the crystal lattice R_e, which gives one the size of the smallest solid particle, i.e. the cluster (see Appendix 4).

The assessment of R_e allows us to find

- the concentration of electric, temperature and mechanical fields near the tops of the microtips on the surface of electrodes, as well as the distribution of polarization there;
- the eigenfrequency of the oscillations in the electron-phonon system, which determines the operating frequency not only of diodes and transistors, but also of electrically active structural microdefects. This helps to determine the frequency of injection and plasma chemical methods and impulse electromagnetic fields to eliminate technological defects and in strengthening technologies (used industrially);
- the energy of surface tension in solids (the fundamental characteristic of strength).

The size of the R_e cluster seems to be the most probable minimal size of dislocations and microcracks (in the order of 10 or 100 crystal lattice constants).

Piezoelectric transducers with microprocessors are used in household appliances and other devices [1, 2, 5, 6, 8, 9, 11, 12, 23, 24, 29, 43, 47, 49, 58, 84]. Piezoelectronic computing devices in monitoring and control systems are detailed in [1, 18, 19, 23, 24, 38, 60, 76, 85, 110, 111]. Nondestructive methods of testing materials and constructions are considered in [19, 20, 24, 27, 29, 38, 45, 53, 73, 85, 87, 110, 111]. This book describes how electrophysical and temperature phenomena are connected with mechanical stress-strain state in polar DE. A combination of metal conductivity (determined by the injection of electrons in DE from the cathode) and polarization explains the *unique* properties of *thin-layer MDM structures* used in contemporary *dielectric electronic* devices.

A physical understanding of connected electron and polarization processes in view of the real material structure offers the simplest constructions and technologies [1, 3, 4, 11, 12, 17, 24, 42, 58, 66, 67, 80, 86, 109]. For example, structural defects, deliberately created in polar DE and piezo-semiconductors (micropores, dislocations), increase sensor sensitivity by two orders [36, 88, 89].

The nonlinear dependence of ferroelectric properties on the electric field strength, mechanical stresses and strains and temperature, as well as the *high ohmical resistance of polar DE,* allow one to create unique charge coupled devices (CCD), such as spectacles with built-in 16-gram TVs, electron "noses", which analyse air with great accuracy (identifying about 48 various substances), acoustoelectronic generators, CCD matrixes, programmable microprocessors and single-crystal microcomputers, which require by 2-3 orders less energy than their semiconductor analogs (and operate faster) [1-3, 24, 48, 69, 102].

Electroelasticity relations and basic design of piezoelectronic transducers are discussed in [2, 8, 10, 14-18, 23-24, 30-33, 72, 73].

It is common to use CVC for analyzing the work of electronic devices; in paragraphs 3.1; 3.4; 3.8 we propose another diagnostic parameter, *heating temperature,* which can be precisely and easily measured in the IR range [24, 60].

The non-destructive temperature remote sensing of elements and technological processes in various appliances and devices is considered in [1, 85, 110, 111, 119, 121]. The simplest technique in heat defectoscopy is to register the temperature field on the surface of the tested product (fiberglass, composites) with thermal television equipment or a heat defectoscope, and then make a tomographic computer analysis of thermogrammes [110]. Structural defects are characterized by local temperature changes.

Physico-mathematical models of the techniques for structural defects identification in MDM structures under an external constant electric field in
view of heating temperature are given in paragraphs 3.1; 3.7; 3.8 [58, 65,
87, 119].

For IR remote temperature sensing of fast-moving objects it is necessary
to ensure the displacement of the PZ chopper (or radiation modulator) at
1.0-1.2 kHz. The known piezoelectric bimorph actuators operate at 40 Hz.
However, after $10^5 - 10^6$ deformation cycles the bimorphs split. The construction of a high-frequency piezoelectric actuator in the form of a pre-
bent thin piezoceramic plate with uneven thickness polarization is considered in [121]. This construction can be used as a deflector of laser radiation (e.g., in holographic information recording).

3D electroelasticity relations for PZ ultrasonic receivers are viewed in
[122].

Engineering methods and procedures for calculating the parameters of
stable self-oscillations in electric circuits can be found in [45, 51, 62, 66,
88, 90, 93, 95, 99, 100, 104].

Information measuring systems (IMS) contain automatic frequency and
phase control devices. They are commonly used for autoregulation in radioelectronic systems [99, 102, 103, 105]. Dynamic characteristics of the
systems (in the phase plane), phase control systems with filters, calculating
the boundaries of frequency capture area are covered in [45, 90, 94, 99,
102, 105].

The several parameters of phase synchronization systems (PSS) for
automatic testing must offer *precision, fast response* and *stability* at the
same time, which presupposes *trade-off* decisions. Increasing precision in
two- and three-contour PSS, iteration and adaptive synchronization systems considerably complicates their construction [62, 105-107]. In § 3.5
we use numerical calculation and experimental data to propose a simple
technique that would help (by choosing a polar DE or piezosemiconductor
with a certain electron conductivity σ and relative dielectric permeability
ε_{33}^T characterizing the relaxation properties of the polar DE) to attain
phase autosynchronisation (or frequency capture) in acoustoelectronic
generators (AEG) with one controlled parameter instead of three. This *universal* method can be used in the design of all resonance piezoelectric
transducers (sensors and actuators).

Basic circuit technology of electronic control systems, including coupling sensors with microprocessors, is described in [6, 24, 48, 51, 52, 62,
102]. The construction and circuit technology in [6] are manufactured by
hybrid technology with *standard microcircuits* produced abroad and developed in the 70 s and 80 s of *the previous century,* whereas this book is

mostly concerned with cutting-edge thin film microelectronic technologies and submicron transistors.

Solid-state micro- and nanoelectronics (single-chip computers) makes use of heterogeneous structures, in which *the active element* is *the interface* of *different materials* (metal-dielectric, dielectric-semiconductor, metal-semiconductor). The instability of these devices is caused, among other reasons, by *instable electronic processes* connected with the formation and development of *structural defects* (dislocations, pores and microcracks, or interface surface roughness). The shape and size of pores and rigid inclusions in DE influence the performance of solid-state devices (in view of the injection of electrons from the cathode and "holes" from the anode at thermal and autoelectron emission).

Present-day microelectronic devices work in extreme critical modes, i.e. at the reversible initial stage of electrothermal breakdown (without irreversible destruction of the crystal lattice). The influence of structural defects is *ambiguous*: on the one hand, under certain conditions (§ 3.2.1; 3.4; 3.7; 3.8) minor *random fluctuations* of the electromagnetic field *increase*. Thus, under a constant external electric field in MDM structures we find the autooscillation of the injection current and electromagnetic radiation from the DE into the environment. This forms positive feedback and leads to *irreversible electrothermal breakdown of MDM structures* and insulation in electrotechnical devices. On the other hand, deliberately introduced structural defects make *the sensibility of piezoelectric sensors abnormally high* [36, 70, 80, 88].

The third chapter of this book contains an estimation of the critical electric field strength, at which at the initial reversible stage under a constant external field there are oscillating defects in the microstructure of DE diodes. Therefore, MDM structures can be used as generators, triggers and miniature oscillation microcircuits, including capacities (crystal lattice) and inductivities (formed by spiral-like waves of low-temperature solid-state plasma in electron conductivity channels, or current filaments).

Polarization, electroding, connection of elements in heterogeneous structures are the main technological operations in the manufacture of piezoelectric transducers and solid-state electronic devices. It is necessary to achieve *not only maximal and even DE polarization*, but to maintain *a certain level* at which DE have the best possible temperature and time stability. The geometry of the surface roughness of metal electrodes affects the distribution of polarization along the thickness of MDM structures and the concentration of electric, temperature and mechanical fields (the electric field strength may exceed *by two orders* the average strength, *integral* to the thickness). Local non-homogeneities of these fields have a decisive effect on the electrochemical and physical processes, which determine the characteristics of solid-state electronic devices.

The controlled initial stage of electric breakdown is used to build vacuum luminescent screens and "*quantoscopes*", in which the function of the "target" is performed by DE and semiconductor plates, as well as standard TV vacuum tubes. The laser beam in red light locates cancer cells, whereas the laser in blue light incinerates them.

As we mentioned before, in vacuum thin MDM structures develop *pre-breakdown* phenomena, controlled by an external electric field. For example, the "electrical forming" creates "through" conductivity, electron emission centres and luminescent radiation. This process is used to produce *low-voltage* "cold" cathodes for a *flat* vacuum *luminescent screen*. Electrode roughness is critical in this construction. (In some cases the electric strength of MDM structures is increased with "*island*" metal electrodes). At present injecting low-voltage "cold" cathodes use thin films made of artificial diamond. The main parameter is *the curvature radius of the tops of micropeaks* on the diamond surface. An experimental technique for measuring this parameter is given in § 3.8.

There are various high-energy electrophysical technologies, such as irradiation with high-energy electrons, ion implantation, electron-beam treatment and electroadhesive coupling of different materials (glass-metal, ceramics-metal). For example, the polarization ("charging") of electret polymer films is achieved by electron irradiation or in a gas discharge (plasma technologies). The accumulation of the space charge in electret or oxide films is accompanied by an electric current, DE polarization and electric breakdown. These physical processes are considered in [19, 22, 25, 36, 88, 110, 111].

Due to uncontrollable technological fluctuations, the parameters of solid-state electronic devices do not meet all requirements. First and foremost, this is due to the state of surface and *contact phenomena at metal – DE interface* and to the *electron injection into DE*. The currents of thermo- and autoelectron emission MDM structures, as well as their dependence on the electric field and temperature, have been measured; the nature of injection currents can now be explained by imperfections of the structure. The experiments have shown that *physical models of the ideal crystal lattice are not correct*, if the energy scale of the changes in electronic processes is *about several meV* [20, 22].

Under an external electric field negative charges on the cathode and positive charges on the contacting DE side form a double electric layer with gradients of the electric potential inside, and therefore the electrochemical activity of the materials increases. Atoms in the air oxygen, adsorbed by the DE surface, "entrain" electrons and form a layer of negative ions. This causes the ions of the main material to electrodiffuse towards the DE surface and form chemical bonds with the charged oxygen atoms. Thus, the surface is oxidized.

At "dot" contacts (near micropeaks on the surface of metal electrodes) and around "current filaments" high ohmical DE is subjected to *Joule heating*. Under electrothermal and *thermoelastic mechanical fields* the structure develops defects of submicroscopic/microscopic size. This also gives rise to chemically active centers, which determine solid-state chemical reactions.

Electroadhesive junction of materials requires research into physicomechanical processes at their interfaces. The adhesion of materials is determined by differences in their electrochemical potentials, temperature expansion and Poisson coefficients, moduli of elasticity and contact surface roughness. Polarization and mass transfer along structural defects (diffusion, chemical reactions) form *a thin "interphase" transition layer*, the mechanical characteristics of which considerably differ from the space properties of the contacting materials and largely *determine the strength of the composite on the whole*.

Apart from silicon, microprocessors, information and measurement devices contain gallium arsenide. The microstructure of thin-film systems depends on the temperature of technological processes. The surface morphology of natural and anode GaAs oxide films (grown on plates from p-GaAs by gas phase epitaxy) was studied with an atom force microscope. It showed that the synthesis temperature determines the roughness of the GaAs film. The films grown at temperatures below 200°C had a rather "even" surface. Increasing the deposition temperature up to 300°C forms an island *fractal defect mesostructure* on the film surface [18, 36, 61, 69, 86].

Components of solid-state electronic integral devices (ID) are joined by various technological operations, such as spraying, epitaxy and thermal compression in combination with electric fields (electroadhesion). The process causes preliminary (residual) internal mechanical stresses (IMS), which in SiO_2– Si structures equal approximately $(2 \div 3) \cdot 10^2$ MPa (i.e. approach the permissible maximum). The structural defects and IMS determine ID characteristics, changing their surface and space resistance. Electric and mechanical aging, as well as impurity diffusion, distort the crystal lattice and give rise to IMS.

Information measurement systems are based on optoelectronic devices (planar and fiber lightguides, photo- and light-emitting diodes, optical and laser radiation modulators), manufactured from transparent monocrystals, ceramics and glasses [1, 9, 12, 54, 67, 84, 89]. The Research Centre for

Laser Materials and Technologies (the Institute of General Physics at the Russian Academy of Sciences) has designed a fundamentally new technology for producing high-temperature transparent DE by high-frequency (HF) heating and melting in a "*cold*" container. Using methods of physical chemistry and solid-state physics, the scientists have developed *new techniques for studying the DE structure and calculating the energy evolution of their HF electromagnetic field*. These techniques can be used for the manufacture, polarization and combination of piezoelements, which work at high temperatures, need high dielectric permeability ($\varepsilon_{33}^{T} \geq 2000$), electric durability and quality.

Assembly is the main technological operation in the manufacture of piezoelectronic devices. Elements of piezoelectric transducers are joined by welding and USW sealing. These heating and melting techniques can be used to produce and repair microprocessors in mobile radiotelephones (in which the size of filters and other elements is dozens of times smaller than the soldering drop).

Heating a DE under a HF electromagnetic field

Oxides and glasses are DE; at room temperature their specific electric resistance $\rho_{sp} = (10^{6} \div 10^{18})$ Ohm·cm. It is known that ρ_{sp} drops as temperature rises. It is illustrated in Fig. 28 for some crystal DE [Osiko V.V. Laser Materials. Moscow: Inst. General Physics. 2002. (in Russian)].

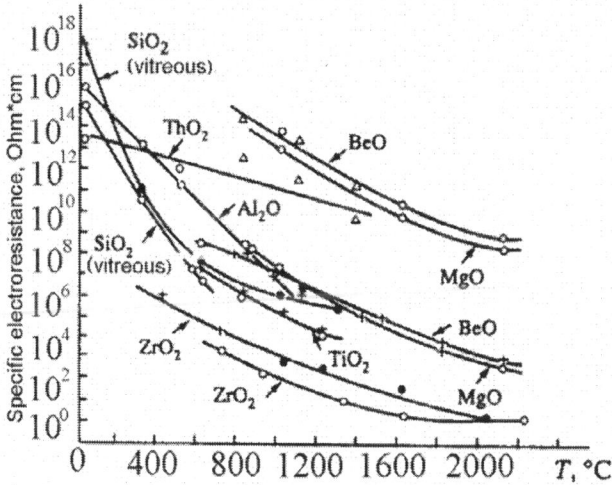

Fig. 28. Dependence of specific electroresistance on temperature for some oxides

The specific resistance of oxides ρ_{sp} is $3 \div 10$ Ohm·cm. On the basis that ρ oscillates in the range of $0.1 \div 3$ Ohm·cm, let us calculate the frequency of a HF device, which would allow *effective heating of the DE*. The effective inductive heating of the material in a cylindrical inductor, if the HF current penetration depth does not exceed the half-diameter of the heated body, is

$$\delta = \sqrt{\frac{2\rho_{sp}}{\omega\mu_m}} \leq \frac{d_M}{2}, \qquad (1)$$

where δ is the HF current penetration depth; ρ_{sp} is the specific resistance of the heated body; d_M is the diameter of the heated body; ω is the circular frequency; μ_m is the magnetic permeability.

The electric performance (η) of the system "inductor – heated material" depends on the correlation of the diameter of the material d_m and the penetration depth of the HF field δ. The nature of this dependence is shown in Fig. 29. The minimal frequency of the alternating current that allows efficient heating is

$$f_{min} = \frac{(4 \div 9)\rho_{sp} \cdot 10^8}{d_m^2 \mu_m}. \qquad (2)$$

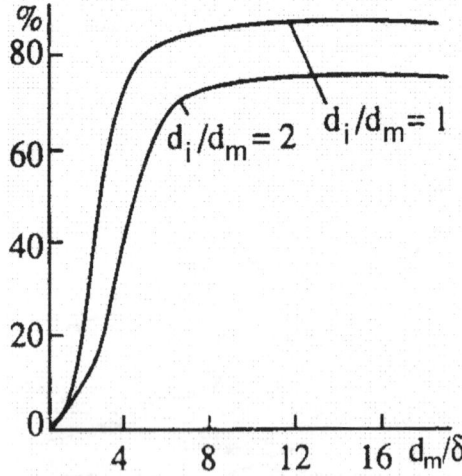

Fig.29. Dependence of the HF heating performance on the correlation of the inductor and heated body diameters and the penetration depth of HF current [83]

For no less than 50% generator performance, we choose $\dfrac{d_M}{\delta} = 4$. Then, from (2) for the lower frequency limit at $\rho=0.1$ Ohm·cm, $d_m=5$ cm, $\mu_m = 1$.

$$f_{min} = \frac{4 \cdot 0.1 \cdot 10^8}{5^2 \cdot 1} = 1.6 \text{ MHz.}$$

For the higher frequency limit at $\rho_{sp} = 3$ Ohm·cm, $d_m = 5$ cm, $\mu_a = 1$.

$$f_{min} = \frac{4 \cdot 3 \cdot 10^8}{5^2 \cdot 1} = 48 \text{ MHz.}$$

These results correspond to experimental data for polyamides, polyethylene, and cellulose. The materials were made more durable by the plasmochemical technique, at the frequency $f = 13.5$ MHz.

For direct heating of DE, two HF generators were calculated and built, with the power $P_1 = 18$ kW, $P_2 = 60$ kW and the frequency $f = (2+10)$ MHz. The generator frequency was tuned mostly by the circuit capacity.

The primary goal in calculating and building the HF generator was to make the circuit simple and reliable. The anode and net circuits were joined through the anode-net capacity of the generator lamp. There are no known circuits for frequencies below 10 MHz. It is shown in Fig. 30.

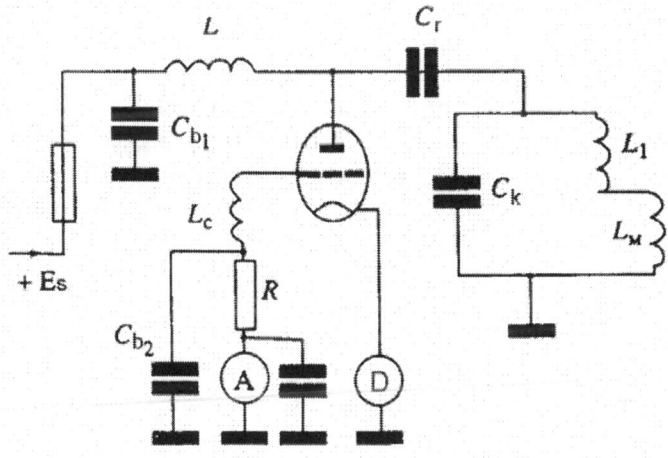

Fig. 30. Layout of the HF generator for direct heating of DE

The use of MW electromagnetic fields for baking ceramics from powdered oxides of heavy metals like $PbZrTiO_3$ (the most commonly used PZT-4 is based on a solid solution of titanium oxides TiO_3 (Fig. 28) forms an ultradisperse structure, which gives high quality and a 1.8-2.2 times increase of piezoelectric constants. *Simultaneous application of a constant external electric field and a HF electromagnetic field, as well as USW oscillations, is used to polarize piezoceramics.*

Plasmochemical techniques and HF electromagnetic fields are used to increase the durability of metals, semiconductors and DE. The main technological parameter of the process, *the frequency of the HF electromagnetic field,* may be determined from formula (2).

Servicing microprocessor control systems for automobile engines is considered in books and technical manuals [1, 8, 12, 18, 24]. Servicing household appliances requires a quantitative evaluation of residual technological mechanical stresses and strains, which occur in elements of control systems and microcircuits. The structure, repair technology and application of PZ transducers are considered in [1, 23, 24].

Silicon-on-sapphire (SOS) structures are the basis of integral *microcircuits,* stable towards destabilizing factors. Because of *crystallographic discrepancy* and *different thermomechanical properties,* SOS compositions have high residual technological mechanical stresses and structural defects. Improving heteroepitaxial SOS structures is of practical importance. If 0.6 μm silicon films (deposited by monosilane pyrolysis on 540 μm-thick sapphire plates Al_2O_3) are subjected to γ-radiaton, the concentration of structural defects is reduced. This is caused by elastic waves, which appear in the sapphire as a result of ionization processes.

For example, the effect of γ-radiation is similar to that of the external electromagnetic impulse and constant fields. Under the conditions determined in Chapter 3 (§ 3.4; 3.7; 3.8) and the Appendix (3, 4), this can heal technological defects. On the other hand, *high-energy radiation* may be used to fabricate a regular system of microdefects, which would increase the electric durability of MDM structures and improve the characteristics of solid-state structures [36, 88].

Structural imperfections account for the uneven distribution of electric mechanical fields in heterogeneous dielectric structures. The Appendix (4) contains formulas for calculating the effective electroconductivity of porous DE and conductors, which correspond to the results obtained from the perturbation theory [70].

The applied theory of autooscillations in radioelectronic devices and *phase synchronization systems* (PPS) are considered in [62, 66, 93, 96, 100]. In PPS the low-frequency filter in the error signal circuit works with microcomputers. The universal method of quantization by count groups can be used not only in synchronization systems, but also in other automatic control systems (ACS) [107]. PSS are used in demodulators and signal filters. Special microprocessor calculators can improve DAC characteristics and create new ACS devices. There are techniques for analyzing digital systems and for replacing an original digital system by an analog model [100, 102].

A digital system with a microcomputer may be equivalent in its dynamic characteristics to an analog delay system. The dynamic characteristics of a PSS with a damping device (DD) are better than those of a system without a DD. A microcomputer can be a digital filter and *a phase detector* (PD). To measure phase difference, it requires 30 counts within the input signal period (no more than 1% in error) [102].

A logical development of negatronics (Chapter 4) is to design neuristor circuits with feedback, which contain DE structures with negative resistance (with internal positive feedback). MDM structures with an S-type CVC are used in photoreceivers, high-speed LSI memory cells and logical circuits with feedback.

Digital *programmable automats in ACS* make them *adaptive*. Special microprocessors are used in computing systems that can work with incomplete information, i.e. robust systems [94].

Historically, the quantitative integral evaluation of polar DE is based on *electric induction, i.e. the surface density of the charge* D_z [C/m^2]. It is conventionally understood that the properties of electrets depend on *the strength of the electrostatic field* in the environment near the surface. However, *the external field strength cannot* characterize either the internal state of the electret, or its local polarization, or the distribution of the space electric charge near the elecret surface.

Physical properties of electrets depend on *electric dipoles* (domains), rigidly mounted and forming an "immobile" crystal lattice (for ion crystals) and *electric charges* (free and bound). An electret does not always contain *static polarization*. It can be obtained by injecting charges into the DE and their "entrainment" by the "traps". The role of "traps" in polymers can be played by structural defects (irregularities of molecular chains), and macroradicals of molecules, which have a high electron affinity. Reference sources do not always cover these fundamental differences [1, 3], which leads to an incomplete and erroneous understanding of physical processes vital for developing *the technology of electrets in solid-state electronics*.

The most general physical model presupposes polarization with *the surface density* P_e and a additional charge, distributed with a *certain space density* P(z) along the electret thickness. One should take into account the unevenness of polarization along the DE thickness and also of the electric, mechanical and temperature fields concentrated near structural imperfections.

Measuring the uneven distribution of polarization along the thickness of thin-layer MDM structures is technically rather difficult. An experimental technique for determining it in a thin piezoelectric plate can be found in § 3.2.1.

The sandwich piezotransducer is considered in [Shuyu L. Optimization of the performance of the sandwich piezoelectric ultrasonic transducers. *JASA*. 2004. Vol. 115. N 1. P. 182-186]; [1, 8, 18, 23, 30, 39].

The most vivid results of reconstructive computing X-ray tomography have been obtained in the medical field. The first tomographs were used to diagnose human brain disorders, and later in the abdomen. Standard tomographs have a resolution of fractions of millimeter. Microelectronic devices can be controlled at a resolution of ~10 μm.

The principles of tomography found application in the development of ultra-high resolution aerials (radio-relay stations with synthesized aperture). [Peterson W.W., Weldon E.J. Error-Correcting Codes. Cambridge, Massachusetts: MIT Press. 1972; Clark Q.C.Jr., Cain J.B. Error-Correction Coding for Digital Communications. NY-London: Plenum Pr. 1982].

A helicopter flying laboratory realizes a complete digital processing of the radio location signal in real time. For the first time it was possible to achieve the adaptation of Radon's integral transformation for noncoherent information and image synthesis [66].

The fundamental issues of modern-day physics are largely nonlinear. Solving them, researchers hope to create a physically coherent "world picture" [7, 25, 45, 62, 75, 90-107, 113, 115, 116].

Supplementary sources of information

Results of experimental research into electroadhesive compounds are given in [120]; [Jonsen A., Rahbek K. J. of Intern. Electric Eng. 1923. Vol.61, p.713-725; Winslow V.M. J. Appl. Phys. 1949. Vol.20, p.1337-1140; Schaffert R.M. Electrophotography. London, New York: The Focal Pr. 1965; Wallis G. Electrocomponent Science and Technology. 1975. Vol.2. No.1, p.45-53; Bischof C., Possart A. Adhesion – Theoretische and Experimentale Grundlagen. Berlin: Akademie-Verlag. 1983].

"In 1923 cohesion of solids at application of external electrical field was disclosed by Rahben and Jonsen. Then in 1960 method of soldering the ceramics to ceramics and the glass was suggested. There exist the contstrains between the value of electrization and surface tension (i.e. the work adhesion) according to Dupre-Young equation" [Panasjuk J. J. *Exper. And Theor. Phys. JETP.* 1939. Vol.9, p.1245 (in Russian, translated into English)].

Measurement techniques and devices, systems for failure diagnostics in electric circuits and equipment, as well as the circuitry of analog electron devices, integrated circuits, failure-proof microcontrollers, measurement-computing circuits and devices are considered in [114-126].

The modeling of electronic and electrotechnical devices is performed with standard programmes like MS EXCEL, etc. The applied theory of information transmission and processing is expounded in [35, 38, 45, 51-58, 64-69, 85, 87, 108].

Repairing methods of electronic equipment without circuits and practical information about electronic devices are given in [119]; [Bogomol'nyi V.M. Resonance calculation for electrothermal damage of metal-insulator-metal (MIM) structures. *Measurement Techniques.* 2000. Vol.43. No.6. p.538-543; Bogomol'nyi V.M. Calculation of critical parameters of self-exciting processes in the electrothermal destruction of dielectric diodes. *Measurement Techniques.* 2000. Vol.46. No.11, p.973-981].

The physical essence of electrophysical processes, constructing automatic devices, information-measurement and computer devices (including nonlinear electromagnetic phenomena in ferroelectrics and ferromagnetics) are considered in the light of technical electrodynamics in the monograph [46-55, 108, 112].

"Closed form analytical equations are tool for optimizing of a PZ actuator for micropump of the base of three layer composite structure (with passive plate and the bonding thin layer). Based on proposed model the effects of main parameters and nondimensional variable groups on the PZ actuator performance have been investigated. The obtained results shows excellent agreement with experimental data "[S. Li, S. Chen. Analytical analysis of a circular PZT actuator for valueless micropumps. // Sensors and Actuators A. Physical. 2003. Vol. 104. N 2. P. 151-161].

. "Porous ceramics of PZT were prepared by sintering powder compacts consisting of PZT and stearic acid powders in an air atmosphere. Stearic aced was added as a poreforming agent (PFA). The dielectric, elastic and piezoelectric properties of uniform by porous PZT ceramics were investigated as a function of the porosity volume fraction.

"The electric-field-induced bending displacement characteristics of the PZ actuator were measured [Jing-feng Li et al. Fabrication and evolution of porous piezoelectric ceramics and porosity-graded piezoelectric actuators. *Journ. of the Amer. Ceramic Soc.* 2003. Vol. 86. No. 7, p. 1094-1098].

"The investigation focuses on the strain transfer mechanism between the active PZ layer and the host structure, the stress concentration is involved, and the influence of the geometrical and elastic parameters of the adhesive layer on static response of the PZ actuator. The numerical analysis is based on the high-order approach and uses 2D elasticity equations of the adhesive layer" [Rabinovitch O., Vinson J.R. Adhesive layer effects in surface mounted PZ actuators. *Journ. of Intelligent Material Systems and Structures.* 2002. Vol. 13. No. 11, p. 689-704].

Experimental data on electron emission at brittle destruction of ion crystals and of the kinetics of cracking in the external electric field have shown that the *charge* density at the top of cracks at destruction is by 2-3 orders bigger than the density of residual charges in the volume. Experimental data of the kinetics and concentration of the charge at the top of a moving crack were obtained in [Gershenzon N.I. et al. Electromagnetic radiation of the crack top at destruction of ion crystals. *Dokl. Akadem. Sci. USSR.* 1986. Vol. 288. No. 1, p.75-78; Surkov V.V. On electron emission at crystal DE destruction. *Sov. Phys. – Techn. Phys.* 1986. Vol.56. No.9, p.1818-1820; Molotsky M.I., Malyutin V.B. Energy spectrum of mechanoelectrons. *Sov. Phys. Solid State.* 1983. Vol. 25. No.10, p.2892-2895; Molotsky M.I. *Sov. Phys. Solid State.* 1976. Vol. 18. No.6, p.1763-1764).

The radiation can be explained within a dislocation model. Mechanical stresses near the crack top (exceeding the yield stress τ) generates and moves along <110> edge dislocations. The break of charged dislocations off *the Debye-Huckel screening spheres* separates charges so that the crack top moves along with a *negative charge*, whereas the equal positive charge is at the start of the crack. Let us calculate the charge. *Dislocation velocity* V_d is determined by the local value of strain tensor:

$$V_d \cong b\sigma / B,$$

where b is the modulus of the Burgers' vector, B is the value characterizing viscous friction at the movement of dislocations,

$$\sigma \cong \frac{K}{(2\pi)^{1/2}(x^2 + y^2)^{1/4}},$$

where K is the coefficient of stress intensity, x is the coordinate along the movement line of the crack, y is perpendicular to the cleavage plane. If the initial position of dislocation in relation to the movement line is that at all times $V_d \ll V_T$ (V_T is the speed of the crack growth), the shift of dislocation in the stress field is negligibly small. In case of $V_d \gg V_T$ the dislocation is able to move to the distance $y \cong y_c$, where y_c is determined from the equality $V_d = V_T$:

$$y_c \cong \frac{1}{2\pi}\left(\frac{bK}{BV_T}\right)^2.$$

Thus, the presence of a charge in the crack top is ensured by the dislocations formed in the band $-y_c \le y \le y_c$. The linear density of the charge can be calculated by the formula $q = q_0\rho_0 y_c^2 \cong \dfrac{q_0\rho_0}{4\pi^2}\left(\dfrac{bK}{BV_T}\right)^4$,

where q_0 is the linear density of the dislocation charge, ρ_0 is the density of dislocations. The obtained formula can be applied at $y_c \le y_0$, where y_0 is determined from the expression $K(2\pi y_0)^{-1/2} = \tau_T$; at $\tau_T = 5 \cdot 10^7$ dyne/cm^2, $y_0 = 0.05$ cm. At $\rho_0 = 10^8$ cm^{-2}, $b = 2.8$Å, $K = 3 \cdot 10^7$ dyne/cm$^{\prime 2}$, $V_T = (0.5 - 1) \cdot 10^5$ cm/s, $B = 2.5 \cdot 10^{-4}$ g/cm·s, $q_0 = (10^{-2} - 10^{-1})e/a$, a is the lattice constant, e is the electron charge, we obtain $q = 10^{-1} - 10^{-3}$ SGS units/cm, which is in accordance with the charge determined experimentally. When velocity is reduced, the linear charge density grows to maximum $q_{max} \cong q_0\rho_0 y_0^2 = 5(10 - 10^2)$ SGS units/cm and is maintained to a total stop. Let us estimate the power of electromagnetic radiation P, which is generated by the charge moving with the crack top:

$$P \cong q_{max}^2 a^2 / c^3 \cong 10^{-6} - 10^{-2} \text{ erg/s},$$

where $a = 10^{11} - 10^{12}$ cm/s^2 is the acceleration of the top slowdown, obtained experimentally. It is conveniently believed that the electromagnetic radiation of an opening crack is caused by hastened movement of charged crack edges. It was found that the radiation power of the charges moving with the crack top P_v considerably exceeds the radiation power of the moving crack edges P_b.

The effect of stepwise resonance magnetization of ferroelectrics was discovered in 1917. It is now used for non-destructrive testing of structural defects in conductive materials by their interaction with a stochastic EMR (*induced by magnetization* jumps). Thus, one can use the technique based on the *Barkhausen effect* (BE) for testing *non-magnetic metals* (aluminum and titanium alloys) [Vengrinovich V.L., Busko V.I. Magnetic noise technique for testing chemical composition of ferromagnetic alloys. *Defectoscopy*. 1982. No.2, p.36-44 (in Russian)].

The Barkhausen effect is used for measuring *velocity, acceleration, displacement and microstrains* [Lomayev G.V. et al. Barkhausen effect in non-destructive testing. *Defectoscopy*. 1984. No.3, p.54-70]. This effect represents stepwise changes in magnetization (*Barkhausen jumps – BJ*), caused by various external influences on the ferromagnetic (*heating, strain, current conductivity and at slow changes of the magnetic field*). BE includes steps of magnetization at *martensite transformations and interrupted motion of cracks in a fractured solid body*. Stepwise repolarization in ferroelectrics and ferroelastics are also BE.

An experiment with *thin carbon steel wire* registered $2 \cdot 10^6$ BJ at a lengthening of 1 mm (!). This explains the high sensibility towards the magnetic field (10^{-4} A/m) [Rosenblat M.A. Magnetic sensors. Present-day situation and development tendencies. Automatics and telemechanics. 1995. No.6, p. 3-55].

The energy that escapes at BJ is enough to be registered by standard radiotechnical techniques, such as induction, capacity (dielcometry), piezoelectric and galvanomagnetic. Contactless information reading is possible at the distance of 1 m; this distance increases with the increase in the sample volume and decrease in the BJ length (within $10^{-4} - 10^{-6}$ s for conducting metals, 10^{-7} s for ferrites and thin films).

The random character of the BJ flow does not introduce errors into measurements. Because random values are added, it is possible to tune off the determined component by algebraic subtraction. This unexpected positive effect is used in the remagnetization by *the gradient and rotating* fields, at which the average magnetic coupling is zero. The output signal contains only the measurable random component.

The electromagnetic signals of BJ are both "*surface*" and "*explosive*". It is explained by eddy microcurrents and BJ locality.

At first the whole energy flow is displaced by an eddy current on the surface and closed through the air medium. After that, it slowly "deposits" on the surface of the ferromagnetic. After the initial transitional process, BJ go through the ferromagnetic. This peculiarity helps to realize various constructions of BE-based sensors. There is a "*chemical BE*", which makes the speed of chemical dissolution of a ferromagnetic depend on BJ intensity.

Negative differential resistance (NDR)

We have studied stationary modes in the circuit containing an S-type NDR element. The circuit under study is shown in Fig.33. It contains a series of an NDR element, the load resistance R, the voltage source E. The mathematical model of stationary modes of the system is generally a system of nonlinear algebraic equations of the form (Fig.34)

$$F_j(U_j, \Lambda_j) = 0, \qquad j = 1, \ldots, m,$$

where U are the falls of voltage on the elements in the equivalent circuit of the NDR device, Λ_j is the set of parameters describing the passive and active elements. This model allows one to join currents and voltages in the equivalent circuit of the NDR element and the external circuit.

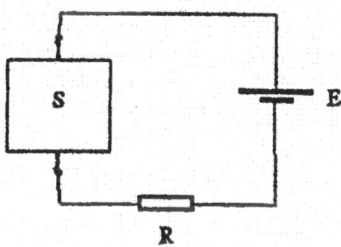

Fig. 33. Layout of stationary modes of NDR elements: S is the NDR element; R is the load resistance; E is the voltage source

These devices have an NDR zone on their CVC (Fig. 35, curve A). The current jump (Fig. 35, curve B) is realized in an MDS-structure without a barrier in the substrate.

S-type NDR elements were represented by a thyristor (its equivalent on two bipolar transistors) and multi-layer structures (Fig. 34, A, B, C correspondingly).

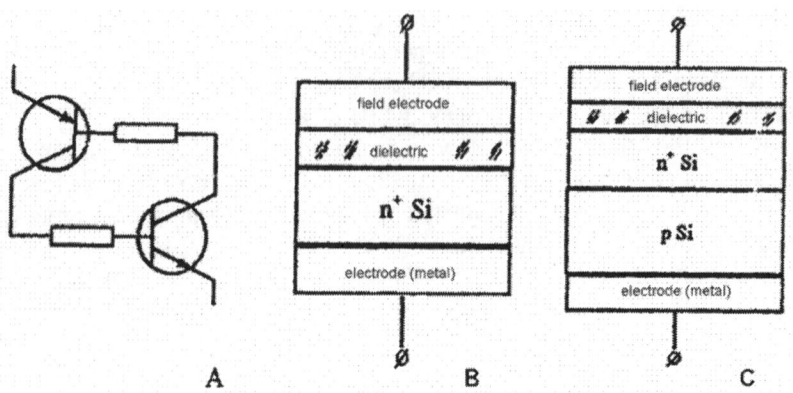

Fig. 34. NDR elements under study: A is the thyristor equivalent; B is the MDS-structure with a uniformly alloyed substrate; B is the MDS-structure with a non-uniformly alloyed substrate

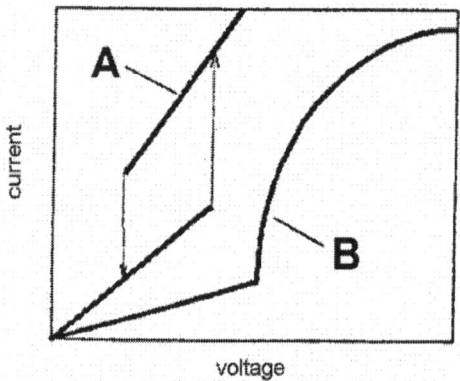

Fig. 35. Idealised CVCs for multi-stable systems

To model the thyristor, we have used a well-known relation

$$I(U_s) = I_0 = \frac{I_{k1} + I_{k2}}{1 - (\alpha_1 + \alpha_2)}.$$

Multi-stable semiconductor devices with a S-type CVC can preserve information, transform analog signals into discrete signals (key elements, shift registers, parametric control generators). Their work can be analysed by the equivalent circuits technique (see Fig. 34).

Electromotive force (EMF) at shock wave distribution through DE

We consider the problem (approximately and precisely) of the current in the circuit of a short-circuited condenser under shock compression. We supposed that when the shock wave travels through DE, there is a surface electricity charge at the wave front. The material before the front was considered an insulator, and the material after the front was a conductor. The precise analytical solution was obtained by supposing the absence of material compression and changes in dielectric permeability behind the front of the shock wave [Zeldovich Y.B. Electromotive force arising at shock waves in DE. *J.Exper.Theor.Physics (JETP)* 1967. Vol.53. No.117, p.237-243]. Academician Y.B.Zeldovich designed the atomic bomb and missiles and rockets for Katyusha mortar.

The circuit of a short-circuited condenser with a solid DE under shock compression develops a current [Ivanov A.G., Lisitsyn Y.V., Novitsky E.Z. Problem of DE polarization at shock load. *J.Exper.Theor.Physics (JETP)* 1968. Vol.54. No.1].

At the shock wave front (SWF) due to instantaneous compression and non-symmetry (uncompressed material before SWF and compressed after SWF) there is a surface charge with a density σ. This assumption is equivalent to the supposition about the *polarization of a DE by SWF*, where the density of the polarization current i is (Fig.36)

$$i = \alpha\kappa T[\kappa T + (1-\kappa)t]^{-2}, \qquad (1)$$

where $\kappa = \varepsilon_2\delta/\varepsilon_1$, ε_1 and ε_2 is the dielectric permeability of the material before and after SWF, δ is the compression, $T = a/D$ is the SWF run time at the speed D of the initial thickness of the dielectric a, $\delta = i_{max} a/D$

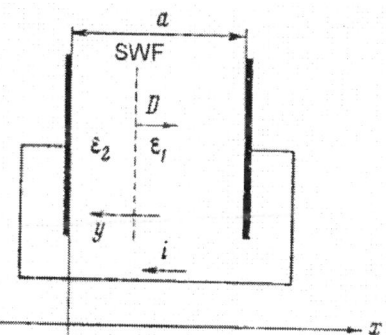

Fig. 36. Layout of a circuit of a short-circuited condenser with a DE. x is the immobile coordinate axis, y is the coordinate axis connected with SWF

Approximate solution. Quantitative differences of conductivity will occur if $\Theta \ll T$, where Θ is the characteristic time for charge drain in the condenser due to conductivity: $\Theta = \rho\varepsilon_2 / 4\pi$ (ρ is the specific resistance behind SWF.) In this case there will be a layer of compensating charge with a volume density v behind SWF.

In a natural system of coordinates, connected with SWF, when the coordinate y is marked in the direction of the material flow, the current i apart from the component proportionate to the field E will contain a component which describes charge transfer with the material

$$i = E/\rho + vD/\delta. \qquad (2)$$

The condition of the stationary state ($i = 0$) allows one to connect E and v:

$$v = -E\delta(D\rho)^{-1}. \qquad (3)$$

Equation (3) with the Poisson's equation

$$\varepsilon_2 (dE/dy) = 4\pi v \qquad (4)$$

gives

$$dE/dy = -4\pi\delta/D\rho\varepsilon_2. \qquad (5)$$

Integrating (5) at $E|_{y=0} = E_0 = 4\pi\sigma\varepsilon_2^{-1}$ allows one to determine

$$E = E_0 \exp(-y/y_0), \qquad y_0 = \rho D\varepsilon_2 (4\pi\delta)^{-1} = \Theta D\delta^{-1},$$

since in the layer y_0 behind SWF there is a step of the potential of the order μ, where

$$\mu = \int_0^\infty E\,dy = E_0 y_0 = D\rho\sigma\delta^{-1}. \qquad (6)$$

If the circuit is short-circuited and ρ is small, μ drops on the layer $a - x$ and creates a field (x is the path made by SWF on the DE) $E = \mu/(a - x)$, to which corresponds the charge $S = \varepsilon_1\mu[4\pi(a - x)]^{-1}$ and the density of the current in the chain

$$i = dS/dt = \varepsilon_1\sigma\rho D^2 \cdot 4\pi(a - x)^{-2}\delta^{-1}. \qquad (7)$$

It follows from (7) that $i \to \infty$ at $x \to a$.

In fact, $\rho \neq 0$ and the current contains a load resistance R. That is why formula (7) is limited by the condition that at $x \to a$ the potential difference on the layer of compressed material behind SWF does not exceed μ

$$\rho a i/\delta \leq D\rho\sigma/\delta, \qquad (8)$$

or $i_{max} = D\sigma/a$, i.e. i_{max} does not exceed i, which is determined discounting either conductivity or relaxation. Solving together (7) and (8), we find the applicability of formula (7):

$$a - x \geq (aD\rho\varepsilon_1 / 4\pi\delta)^{1/2} \approx (ay_0)^{1/2}.$$

Accordingly,

$$V_{max} = i_{max}R = D\sigma R / a.$$

Formula (7) also calls for correction in the beginning of the process. Indeed, at $t = 0$ in a real circuit $i = 0$, i.e. the i from (7) *does not appear instantly*. It must be noted that apart from the time Θ, the time of the circuit relaxation $\tau = RC$, where C is the initial capacity. Depending on $\tau <> \Theta$ at the start of the process there may be different curves $V(t)$.

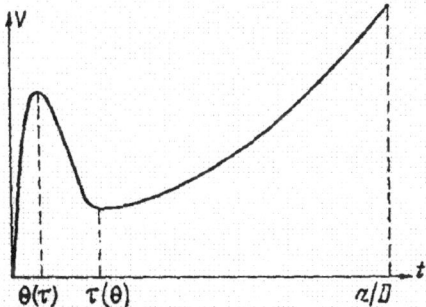

Fig. 37. Fall of voltage on load and its dependence on time

The dependence of $V(t)$ in Fig. 37 is similar to N-shape CVC, characteristic of *autogenerators* (tunnel and Gunn diodes). The physico-mathematical model by Academician Y.B.Zeldovich explains the appearance of spontaneous current autooscillations after the shock of a DE. In piezoelectrics this effect is used in detonators, car spark-plugs, piezolighters and sensors.

There is a vortex flame wave in cylinders of internal combustion engines. The whole pressure of the gaseous combustion products is applied at one point near the perimeter of the piston, which locks the piston, increases fuel consumption and reduces engine power two-fold. These faults could be eliminated by additional polarization of the gas mixture in the process of inflammation.

Piezoceramic spark-plugs create a spark on their own and may be used for automatic (spontaneous) inflammation of the gas mixture.

The physical model of EMF appearance and DE polarization of the gas mixture in the engine cylinder at impulse load can be used to calculate the parameters of an original silicon spark-plug, which eliminates the locking of the piston in the cylinder and allows one to increase the engine power by several times (and noticeably save petrol). The analysis of the vortex movement of the mixture has shown that the combustible mixture should be ignited in *the central part of the cylinder*. It requires a cone screening attachment.

To increase the power of the spark before inflammation, one creates an additional *capacity* (a silicon cylinder with an area of 33 cm^2). The spark-plug is a multi-layer construction of internal and external electrodes with a high-temperature "*silicon condenser*" in between. The calculation methods of the stress-strain state of this cylindrical multi-layer shell are given in the Appendix.

The plug can work as a piezolighter (shock in a polar DE leads to EMF) and create very high voltage.

The theory of electric processes on the interface of dielectric media is expounded in [Yemets Y.P. Electric forces on the interface of DE media. *Journal of Applied Mechanics and Technical Physics*. 1993. Vol. 34. No.4(200), p. 14].

Temperature in the cylinder of a car engine rises, hence spark-plugs should consist of high-temperature ceramics, e.g. porous ceramics. The thermopolarization effect is a linear response of electric polarization to temperature gradient (Gurevich V.L. On electrothermal effect in crystal DE. *Sov. Phys. Solid State*. 1981. Vol.23, p.2357-66; Gurevich V.L., Tagantsev A.K. To the theory of thermal polarization effect in centre-symmetrical DE. *Letters to JETP*. 1982. Vol.35, p. 106-108).

Temperature gradient in solid PZ leads to deformation gradient, which causes polarization in DE (with any type of lattice symmetry) (Matskevich V.S., Tolpygo K.B. Electrical, optical and elastic properties of diamond-like crystals. *JETP*. 1957. Vol.32, p.520-525; Kogen Sh.M. Piezoelectric effect at nonuniform strain and acoustic dissipation of current carriers in crystals. *Sov. Phys. Solid State*. 1962. Vol.4, p. 1765-77; Indenbom V.L. et al. Flexoelectric effect and crystal structure. *Crystallography*. 1981. Vol.26, p.1157-1162).

Oscillations of transparent solid DE at optical radiation

Let us consider a DE with a flow of optical radiation in the direction of the axis x. In this case the heat conductivity equation has the form

$$\rho C \frac{\partial T}{\partial t} = \frac{\partial}{\partial x}\left(\kappa \frac{\partial T}{\partial x}\right) + I\beta, \qquad (1)$$

where ρ is the density, C is the heat capacity, T is the temperature, κ is the heat conductivity coefficient, I is the intensity of the heat flow, β is the absorption coefficient. In equation (1) κ and β depend on temperature [Lysikov Y.I. On the possibility of oscillations when heating a transparent solid DE with optical radiation. *J. Appl. Mech. Techn. Phys.* 1988. No.4(146), p.56-59; Anisimov S.I. et al. Effect of electron heat conductivity on thresholds and development dynamics of DE breakdown. *Quantum Electronics.* 1981. Vol.8, No.8(110); Gershenson N.I. et al. Electromagnetic radiation of the crack top at the destruction of ion crystals. *Dokl. Akadem. Sci. USSR.* 1986. Vol.288. No.1, p. 75-78].

Heat flow I slightly changes with coordinates, which is obviously far from breakdown (intensity changes are relatively small due to absorption at characteristic sizes). The dependence of β on temperature is similar to the one commonly used for nonlinear models

$$\beta = \beta_1 \exp\left\{\frac{E_0 + \gamma\sigma_{xx}}{k_* T}\right\}, (2)$$

where E_0 is the value proportionate to the width of the prohibited zone, σ_{xx} is the component of the elastic stress tensor; γ is the constant coefficient of proportionality. The presence of σ_{xx} in (1) corresponds to the view of dependence of the prohibited zone width on elastic stresses. Equations of the elasticity theory lead to

$$\rho\frac{\partial^2 u}{\partial t^2} = -K\alpha\frac{\partial T}{\partial x} + \left(K + \frac{4\mu}{3}\right)\frac{\partial^2 u}{\partial x^2}, \qquad (3)$$

where $u = u(x,t)$ is the displacement of solid points; K and μ are the coefficients expressed through the Young's modulus E and Poisson's σ by formulas $K = E/[3(1-2\sigma)]$, $\mu = E/[2(1+\sigma)]$ and α is the coefficient of volume temperature expansion. For the value σ_{xx} we obtain

$$\sigma_{xx} = -K\alpha(T - T_0) + \left(K + \frac{4\mu}{3}\right)\frac{\partial u}{\partial x}. \qquad (4)$$

Let us linearize equations (1), (3) against the initial uniform distribution and determine the conditions of oscillations.

We linearize by expanding variables at a point in time by the deviation of temperature $T - T_0$ from the relatively correspondent value T_0. The zero level of elastic stresses corresponds to T_0. The initial distribution of all values is uniform, without dependence on x. Thus, the coordinate derivatives are the values of the first order of magnitude. Because of that the coordinate dependence x is negligible, since this account corresponds to the second order. We write the absorption coefficient as

$$\beta = \beta_0 \exp\left\{-\frac{E_0 + \gamma\sigma_{xx}}{k_*T} + \frac{E_0}{k_*T_0}\right\}, \qquad (5)$$

which corresponds to the definition of β_0 as the absorption coefficient at $T = T_0$. Expansion (5) leads to

$$\beta = \beta_0\left\{1 + \frac{(T-T_0)}{k_*T_0^2}(\gamma K\alpha T_0 + E_0) - \frac{\gamma}{k_*T_0}\left(K + \frac{4\mu}{3}\right)\frac{\partial u}{\partial x}\right\}. \quad (6)$$

Expression (6) contains a definite form of σ_{xx}, determined by formula (4). Introducing the variable θ by the formula

$$T = T_0 + T_0 t/\tau_1 + \theta \qquad (7)$$

and the constants

$$M = E_0/(k_*T_0), \qquad m = \gamma K\alpha T_0/(k_*T_0),$$

$$\chi = \frac{\kappa_0}{\rho c}, \qquad \tau_1 = \frac{\rho c T_0}{I\beta_0}, \qquad c_*^2 = \frac{K + \dfrac{4\mu}{3}}{\rho},$$

after substituting (7) into (1), (3) we obtain

$$\frac{\partial \theta}{\partial t} = \chi\frac{\partial^2 \theta}{\partial x^2} + \frac{M+m}{\tau_1}\theta - \left(\frac{c_*^2\rho\gamma}{\tau_1 k_*}\right)\frac{\partial u}{\partial x}, \qquad (8)$$

$$\frac{\partial^2 u}{\partial t^2} = -\left(\frac{K\alpha}{\rho}\right)\frac{\partial \theta}{\partial x} + c_*^2\frac{\partial^2 u}{\partial x^2}.$$

The right-hand part of the heat conductivity equation in (8) discounts the value $T_1 = T_0 t/\tau_1$ compared to $T_2 = \theta$. It means that T_2 may noticeably exceed T_1 within the considered time, whereas T_1 will remain essentially small. At the same time, there are no limits on the ratio of change speeds T_1 and T_2. The expansions are applicable at $t \ll \tau_1$ and $\theta \ll T_0$.

Solution (8) can be found in

$$\theta = \theta_0 \exp\{i(kx - \omega t)\}, \quad u = u_0 \exp\{i(kx - \omega t)\}. \qquad (9)$$

After placing (9) in (8), we come to a dispersion equation

$$(\omega^2 - c_*^2 k^2)\left(\omega + i\left(\chi k^2 - \frac{M+m}{\tau_1}\right)\right) = i\frac{k^2 c_*^2 m}{\tau_1}. \qquad (10)$$

An important fact, which follows from (10), is that there can be complex values of frequencies and, therefore, oscillations of the type mentioned can grow exponentially.

Equation (10) is cubic in regard to ω and has three different solutions. Let us analyze the behaviour of solutions, supposing that thermoelastic stresses have neglible influence on the parameters of the solid, i.e.

$m/M \ll 1$. We shall bear in mind the following parameters: $\kappa_0 = 10$ W/(m·K), $\rho = 3\cdot10^3$ kg/m³, $c = 1.3\cdot10^3$ J/(kg·K), $I = 10^{13}$ W/m², $\beta_0 = 1$ m⁻¹, $T_0 = 1000$ K, $E_0 = 60k_*T_0$, $\gamma = 2\cdot10^{-3}k_*T_0/p_0$, $p_0 = 10^5$ Pa, $\alpha = 8\cdot10^{-3}/T_0$, $K = 4.2\cdot10^{10}$ Pa, $E = 7\cdot10^{10}$ Pa, $\sigma = 0.22$, $\mu = 2.9\cdot10^{10}$ Pa. The elastic parameters, density, heat capacity correspond to melted quartz. The absorption and heat conductivity coefficients have an approximate value, which allows for pre-heating of the medium. The intensity is taken as equal to typical values for experimental breakdown of solid DE. At the given parameters $m/M \approx 0.1$, $\tau_1 \approx 4\cdot10^{-4}$ s.

In the extreme case $m/M = 0$ equation (10) gives two types of solutions with $\omega_{1,2}^{(0)} = \pm c_*k$ (acoustic oscillations) and $\omega_3^{(0)} = i(M/\tau_1 - \chi k^2)$ (the entropy mode). The oscillations at ω_3 are a typical mode of heat explosion. The corresponding solution grows exponentially at small values of k and attenuates at big values. At $k = 0$ the characteristic time of the mode growth is expressed by the value $\tau_2 = \tau_1/M$. The relation t/τ_2 determines the satisfiability of the condition $\theta \ll T_0$, mentioned above. Let us note that if we consider the relation $t_*/\tau_2 \approx 1$ the condition for the breakdown threshold by laser radiation (t_* is the impulse length), the threshold intensity will be expressed by

$$I_* \approx \frac{\rho c T_0}{\beta_0 t_*}\left(\frac{k_*T_0}{E_0}\right). \qquad (11)$$

When using relations of type (11), one should remember that T_0 is the temperature at which the speed of heat transfer in medium becomes comparable to the transfer speed of elastic disturbances, and E_0 is the characteristic energy in (2) at $T = T_0$.

The threshold value k_1^2, at which $\omega_3^{(0)} = 0$, equals $M/(\tau_1\chi)$. At the chosen values of the parameters $k_1 \cong 2\cdot10^5$ m⁻¹, $\lambda_1 \approx 3\cdot10^{-5}$ m. Acoustic modes at $m/M = 0$ are not connected with the entropy mode and do not grow.

The dynamic model of a string accelerometer

Acceleration sensors with string resonators, or string accelerometers, are considered the most precise and are used in systems for the automatic control of aerospace objects movement along a given trajectory. They are based on autogenerators with electromechanic resonators, and contain a metallic string which oscillates in a magnetic field. To achieve minimal response time and improve the precision and construction of accelerometers is possible only by detailed research of transition processes and fluctuation phenomena in these devices [Zaitsev V.V. et al. Modelling of autooscillations in a generator wth an electromechanical resonator. *Bulletin of Samara State University.* 2003. No. 4].

An equivalent circuit of a string autogenerator is shown in Fig. 38. An electromechanical resonator (EMR) is in the short-circuited loop of the feedback. The loop contains an active element (AE) with a cubic nonlinearity. If the string is declined off balance $U(X,T)$ at point X at a moment in time T, we can obtain a nonlinear integrodifferential equation with partial derivatives. In dimensionless values this equation has the form

$$\frac{\partial^2 U}{\partial T^2} + 2\delta \frac{\partial U}{\partial T} - \sqrt{1+a(T)}\,\frac{\partial^2 U}{\partial X^2} =$$

$$= 2\gamma B(X)\left(1-\left[\int_0^1 B(X)\frac{\partial U(X,T)}{\partial T}dX\right]^2\right)\left[\int_0^1 B(X)\frac{\partial U(X,T)}{\partial T}dX\right]. \quad (1)$$

In this case $B(X)$ is the dimensionless induction of the magnetic field, $a(T)$ is the acceleration; parameter δ characterizes losses in the resonator, and parameter γ characterizes the depth of the positive feedback.

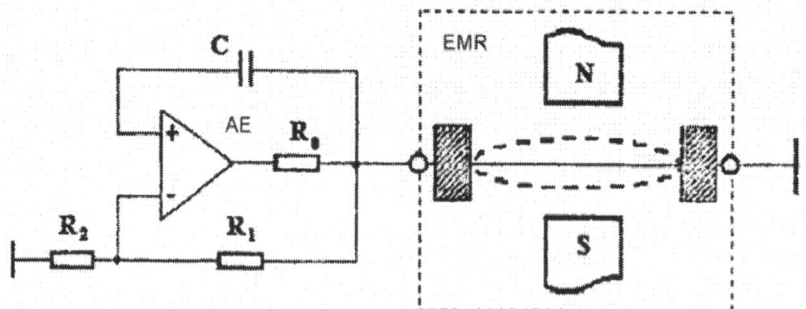

Fig. 38. Layout of string accelerometer

The equation of movement (1) was the basis for two numerical models of the autogenerator. One of them was obtained in the frame of a hybrid computing model, in which the difference approximation of derivatives is made only by the space variable. The other one is a result of the mode expansion of oscillations $U(X,T)$ and represents a system of two Rayleigh's oscillators, which interact through the common feedback circuit. Both models in most cases demonstrate close numerical results. Nevertheless, the model in the form of coupled Rayleigh's oscillators often helps to give them the simplest physical interpretation. For example, it has been shown that a string autogenerator has the effect of "capturing" the harmonics of autooscillations by higher modes of oscillations of the electromechanical resonator.

Appendix

1. Relaxation time of the space charge in a condenser

Under a constant external electric field specific electroconductivity σ is [3, 7, 109]

$$\sigma = en\tilde{\mu}, \tag{1}$$

where e is the electron charge; n, $\tilde{\mu}$ are accordingly the concentration and mobility of electrons ($\tilde{\mu} = v_{av}/E$, where v_{sv} is the average velocity of electron drift, E is the electric field strength).

The drift current density I_{dr} is determined according to the Ohm's law

$$I_{dr} = \sigma E = en\tilde{\mu}E = -en\tilde{\mu}\,\mathrm{grad}\,\varphi, \tag{2}$$

where φ is the electrostatic potential.

After the external electric impulse ceases, polar DE tend towards the equilibrium state.

The equation of electric state, which characterizes the dependence of the DE space charge ρ_3 on the current I_{dr}, has the form [3, 109]

$$\frac{\partial \rho_{\acute{K}}}{\partial t} + \mathrm{div}(\rho_{\acute{K}} v_{dr}) = \frac{\partial \rho_c}{\partial t} + \mathrm{div} I_{dr}, \tag{3}$$

where ρ_c is the density of the space charge, t is the time.

$$\mathrm{div} I_{dr} = -\frac{\partial \rho_c}{\partial t}. \tag{4}$$

Using the Ohm's law (2), we deduce from (3) and (4)

$$\mathrm{div} I_{dr} = \mathrm{div}\sigma E = \sigma \mathrm{div} E = -\sigma \Delta\varphi, \tag{5}$$

where $\Delta\varphi$ is the Laplace operator.

From the Maxwell's equation it follows that [78]

$$\Delta\varphi = -\frac{\rho_c}{\varepsilon_{33}^T \varepsilon_0}, \tag{6}$$

where ε_{33}^T, ε_0 are the relative dielectric permeability and dielectric constant.

From (5) and (6) we obtain

$$\mathrm{div} I_{dr} = \frac{\sigma \rho_c}{\varepsilon_{33}^T \varepsilon_0}. \qquad (7)$$

Equating the right-hand parts of (4) and (7), we have

$$\frac{d\rho_c}{\rho_c} = -\left(\frac{\sigma t}{\varepsilon_{33}^T \varepsilon_0}\right). \qquad (8)$$

Integrating (8), we obtain

$$\ln \rho_c = -\frac{\sigma t}{\varepsilon_{33}^T \varepsilon_0} + C_0, \qquad (9)$$

where C_0 is the integration constant.

Solution (9) at t = 0 gives us that $C_0 = \ln \rho_{c,0}$, где $\rho_{c,0}$ is the density of the space charge at the initial time.

From expression (9) we obtain the formula of the law of the Maxwell's relaxation of the static charge

$$\rho_c = \rho_{c,0} e^{-\frac{t}{\tau_m}}, \qquad (10)$$

where $\tau_m = \dfrac{\varepsilon_{33}^T \varepsilon_0}{\sigma}$ is the relaxation time of the DE space charge.

When a piezoelectric subjected to a single impulse of a mechanical force or temperature, there is an electrostatic field in the polar DE, which decreases with time according to the Maxwell's law of dielectric relaxation (10).

Properties of disordered-phase or defective DE can be controlled, if experimentally studied by *remote pyroelectric analysis* and *dielectric spectroscopy*. A research into low-frequency spectra helped to establish *the universal law of dielectric response* [Jonscher A.K. A new understanding of the dielectric relaxation of solids. *J. of Material Sci.* 1981. Vol. 16, p. 2037-2060]. A study of ferroelectric Langmuir-Blodget films (a copolymer of vynilidenphluoride and triphluoroethylene) demonstrated a radical change in dielectric permeability after introducing small amounts of rhodomine.

2. Maxwell's equations. Joule heat in a cylindrical conductor

The Maxwell's equation system for a conducting solid body without the skin effect has the form [78]

$$\text{rot}\vec{E} = -\frac{1}{c}\frac{\partial \vec{B}}{\partial t}, \text{div}\vec{B} = 0,$$

$$\text{rot}\vec{H} = \frac{4\pi\sigma}{c}\vec{E}, \vec{J} = \sigma\vec{E}, \quad (1)$$

where $\vec{E}$ is the vector of the electric field strength, $\vec{B}$ is the vector of magnetic induction, $\vec{H}$ is the vector of the magnetic field strength, c is the velocity of electromagnetic wave propagation, σ is the conductivity.

Various mathematical techiques to the Maxwell's equations (differential geometry for transient lens synthesis, symmetry and group theory, complex variables applied to frequency, integral operator diagonalization) are formulated in [75].

Equations (1) can be used to solve electrostatic problems on condition that *the frequency of the electromagnetic field ω is small compared to the inverse travel time of electrons in the conductor. The extreme frequencies of electromagnetic wave propagation* correspond to IR range (at the eigenfrequencies of atom oscillations). *In the IR range electromagnetic waves can penetrate metals.* In the optical and radiotechnical ranges electromagnetic waves cannot penetrate metals.

Heating a thin cylindrical wire with the cross-section radius R

The vector of the electric field strength $\vec{E}_z$ and the current vector $\vec{J}$ are directed along the cylinder axis (z)

$$E_z = E_0 \exp(-j\omega t), \quad (2)$$

where j is the imaginary unit, ω is the circular frequency.

An alternating electric field generates a magnetic field, which is determined by the following equation

$$\text{rot}\vec{H} = \frac{4\pi\sigma}{c}E_0 \exp(-j\omega t). \quad (3)$$

In the cylindrical system of coordinates r, φ, z equation (3) projected on the axis z will take the form

$$(\mathrm{rot}\vec{H})_z - \frac{1}{r}\frac{\partial}{\partial r}(rH_\varphi) = \frac{4\pi\sigma}{c}E_0\exp(-j\omega t) \qquad (4)$$

Solving the equation, we find the strength of the non-homogeneous tangential (in the direction of the circumferential coordinate φ) magnetic field

$$H_\varphi(r) = \frac{2\pi\sigma}{c}rE_0\exp(-j\omega t). \quad (5)$$

The amplitudes of the electric and magnetic field strengths are expressed through the alternating current J, taking into account the Stokes' theorem for the circulation of the magnetic field along the contour of the wire with a cross-section radius R

$$H_\varphi(R)2\pi R = \frac{4\pi}{c}J\exp(-j\omega t). \quad (6)$$

From (5) and (6), equating their right-hand parts, we have

$$E_0 = J/\pi\sigma R^2. \quad (7)$$

The average amount of heat radiated within a unit of time per unit of the wire length, in view of (7) equals

$$Q = \frac{1}{2}\sigma E_0^2\pi R^2 = \frac{J^2}{2\pi\sigma R^2}.$$

Under a constant current the latter formula has no multiplier 2 in the denominator, since there is no need for time averaging.

It is known that if an external harmonic field changes at frequency ω, in a very thin cylindrical conductor the current flows only in the thin surface layer of the conductor, the skin layer with a thickness of

$$\delta = \left(\frac{2}{\mu\sigma\varpi}\right)^{\frac{1}{2}},$$

where μ is the magnetic permeability, σ is the electroconductivity.

The numerical analysis of the distribution of the magnetic field and current density in a cylindrical conductor showed that at field reduction there is *"an inverse skin effect"*. The current density is maximal in the centre of the conductor [Kuskova N.I. Inverse skin effect. *Letters Sov.Phys –Techn.Phys.* 2004. Vol.30, No.21, p.59-64 (in Russian), http://www.ioffe.rssi.ru/journals/pjtfl/].

3. Eigenfrequency of low-temperature solid-state plasma oscillations

This section contains a formula for calculating *the eigenfrequency of oscillations* in an electronic subsystem. The *phenomenological* theory discounts the interatomic nature of the material and is based on macroscopic laws and experimental data. In this physical model of electron transport in ion crystals, the electron-phonon system, which is formed by free electrons and bound ("immobile") charges of the crystal lattice, is represented as a *wave* (running from the cathode to the anode in the conductivity zone), the amplitude of which is *counted off the level of the "Fermi's sea"*. The wavelength is determined by the distance between the crests of neighbouring waves [3, 7, 109].

The geometric image of the free electron, the soliton (a solitary wave which does not change shape in motion) seems the most probable. This running solitary impulse exhibits the properties of an elementary particle. The soliton either passes obstacles (local disturbances of the periodical electromagnetic field of the "immobile" crystal lattice) without changing its shape and amplitude or destroys. This idea of how a free electron moves in the ideal crystal lattice was mentioned in 1929 by R. Peierls [Zeitschrift Phys. 1929. Bd. 58, № 1, S.59] *and discussed with Nobel laureate W. Heisenberg.* Later, L.D.Landau developed the view further.

In electron-vacuum tubes, electron-beam TV and computer display tubes the flow of electrons (the "electron ray") is created by "*hot*" exoelectron emission from the surface of the red-hot metallic cathode. Contemporary solid-state counterparts of the electron-vacuum tube differ in that the electron emission from the metallic cathode of semiconductor and DE diodes and transistors occurs *at room temperature*. The thickness of thin-film microelectronic devices varies from one to ten microns; thus, the electric field strength, averaged by the thickness of the MDM structure is hundreds and thousands of times greater than in the electron-vacuum tube (as calculated from the formula $\langle \mathbf{E_z} \rangle = -(2\mathbf{V_0})/\mathbf{h}$, where $\mathbf{V} = \pm \mathbf{V_0}$ are the electric potentials of the electrodes of the DE diode, h is the thickness of the DE layer).

It has been also found experimentally that the *local* electric field strength near *the microtips on the surface of the metallic cathode* can exceed the electric field strength $\langle \mathbf{E_z} \rangle$ (integral by thickness by two orders) [Ridley B.K. *J. Appl. Phys.* 1975. Vol.46. No.3, p.998].

The second difference is that in solid-state analogs of the electron-vacuum tube at the metal-polar DE interface there are *unusual physical phenomena: thermo- and autoelectron emission, acousto-electronic injection* of electrons *in the piezoelectric* [11, 47, 71, 73]. DE have a high concentration of electron entrainment centres ("traps"), about $10^{13} - 10^{15}$ cm^{-3}; that is why they "pump" electrons out of metals and semiconductors and accumulate the charge (the electret effect). The charge accumulated on the traps compensates the "heterocharges" of crystal lattice (polarization) [Ridley B.K. Parametric processes in the acoustoelectric effect. *J. Phys. C: Solid State Physics.* 1973. Vol.6. No.9, p.1605-1614].

The third difference of solid-state plasma from gaseous plasma is that electrons move in a strictly ordered crystal lattice, the fundamental property of which is *translation symmetry.* If the radius of *the spherical zone of the recombination of the heat electron with the positive ion* of the crystal lattice is *comparable to the length of the electromagnetic spherical wave of the free electron*, there is a resonance effect of increasing the velocity of the electron in the autooscillation mode. The *collective* movement of free electrons can be approximated as a movement of *one free electron* in the "ideal" ion crystal (a simplified analog of the electron "ballistic movement", without "collisions" with the atoms of the crystal lattice and its structural imperfections). The electromagnetic wave running along the surface of "the Fermi's sea" is *the enveloping 3-dimensional surface* of spherical zones where heat electrons combine with the ions of the "immobile" crystal lattice. It should be noted that *electron transport* in solid DE occurs along structural defects (dislocations, pores, microcracks) *rather than in the whole volume*. Consequently, the accepted *simplified hydrodynamic model* of the movement of the free electron can be used only for *"ideal" monocrystals* or *thin submicron films*.

The flow of electrons injected from the cathode into the DE and screened by the space positive charge of the crystal lattice ions, forms in local DE zones *quasineutral solid-state low-temperature plasma* (where the concentration of positive and negative charges is approximately the same). Near the microtips on the electrode surface of the MDM structure appear the channels of high electroconductivity, the so-called "current filaments", oriented towards the vector of the electric field strength (towards the thickness of the MDM structure). Because of positive feedbacks of various physical nature (current and temperature) the temperature near the "current filaments" abruptly rises, which increases electron injection into DE and *overheating instability* of the electron-phonon system [28]. As a result of this *instability*, random fluctuations, *brought about by structural imperfections*, increase, and in the *constant external field* we find stable current autooscillations (in some cases accompanied by electromagnetic radiation from the DE into the environment). At harmonic oscillations of the polar DE energy dissipation, which is caused by *dielectric, mechanical and piezoelectric losses, adds this heating to Joule heating.* This rapidly heats the DE near the current filaments to the temperature of mechanical destruction. Thermoelastic stresses in the DE form microcavities (pores) which then contain *low-temperature gaseous plasma.*

Gaseous plasma is principally characterized by *instability*, which under a *constant external field* in a DE *causes oscillations* of the electromagnetic field and current.

The considered nonlinear dynamic phenomena bring S-type zones on the CVC of MDM structures. Near the first cusp of the S-shape CVC there are current oscillations in the external circuit, which contains a DE diode. These physical processes explain the fundamental *difference* between *the CVC of the electron-vacuum tube* and *its solid-state counterpart*, the field transistor (or DE diode). *Gaseous plasma* has *a limit for the emission current*, i.e. there is an extreme constant current called "*the saturation current*" (the Richardson's law). *Solid-state devices do not have such current limits.*

In solid-state physics plasma is understood as *a dynamic system of charged (electrons, ions) and neutral particles, whose concentration is so high that the internal forces of Coulomb's interaction between them are greater than the forces caused by external electromagnetic fields.* This accounts for its *collective response* to any external disturbance. The external electromagnetic field cannot separate positive and negative charges, and therefore plasma "*is a law unto itself*", i.e. *remains independent* of relatively small changes of the environment.

Quasineutrality explains the instability of the balance state of plasma, in which insignificant random fluctuations of the electromagnetic field grow stronger. As a result, if the CVC is S-shape, there are stable autooscillations near the first cusp.

In the early 20th century O.Heavyside discovered that in the Earth's atmosphere at the height of $50 \div 400$ km there is an ionized F-layer, which can reflect radiowaves like a mirror. The F-layer gives stable long distance radio communication [79].

O.Heavyside explained that this layer appears due to atom ionization in upper layers of the Earth's atmosphere by the short-wave Sun radiation in daytime, whereas at night its quasineutrality is maintained by plasmochemical processes with atmospheric ions (atoms), excited by the stored energy of solar radiation. The density of charged particles $10^5 \div 10^6$ cm^{-3} in plasma sharply drops near the layer boundary at the height of 50 km. The temperature of electrons in plasma is approximately $T_e \approx (1 \div 2) \cdot 10^3$K.

Langmuir and Tonks (1923) measured the concentration of oscillating electrons in the electric charge in gases between electrodes. They associated the oscillations to the vibrations in the cellular liquid, the "jelly" which physiologists usually call plasma.

Studying electric arcs in gases between electrodes, Langmuir found that the central part of the charge contains the same number of electrons and positive ions. Near the electrodes this "equilibrium" is broken. The *central part of the gas charge pulsated* like liquid colourless blood plasma, hence the name suggested by Langmuir.

Interactions in plasma are mostly of the Coulomb type (long-distance electromagnetic forces), rather than "interelectron collisions", which play a secondary part. The basic properties of plasma are its *quasineutrality* and *collective response* to external electromagnetic disturbance.

Plasma in semiconductors can be a dynamic system formed by electrons and "holes". *Solid-state plasma, unlike gaseous plasma, can exist in the equilibrium state.*

Oscillating solid-state plasma is formed by injected electrons, free and bound charges and "holes" (limited by a space charge at injection currents). Let us consider the movement of the electron cloud related to the "immobile" periodic system of the crystal lattice atoms. In fact, ions (atoms) perform oscillations (optical and acoustical phonons). The force acting on the electron can be written as [3, 25, 109]

$$m^* \frac{d^2 W}{dt^2} = -eE_e , \qquad (1)$$

e is the absolute electron charge, m^* is the effective mass of the free electron, t is the time, W is the electron displacement under the electric field of the electromagnetic wave. The electric field strength E_e and W change according to the law $\exp(-i\omega t)$. Differentiating by time, we obtain

$$-m\omega^2 W = -eE_e . \qquad (2)$$

Calculating W from (2), we find the dipole moment of the electron Mg=-eW and the dipole moment of the electron gas volume unit, called polarization P,

$$P = -neW = -\frac{ne^2}{m\omega^2} E_e , \qquad (3)$$

where n is the electron concentration.

The expression for the electric induction $D(\omega)$ has the form

$$D(\omega) = \varepsilon_0 \varepsilon_{33}^T(\omega) E_e , \qquad (4)$$

ε_0 is the dielectric permeability of vacuum, $\varepsilon_{33}^T(\omega)$ is the relative dielectric permeability.

The electric induction $D(\omega)$ can be expressed through the electric field and polarization (in SI units)

$$D(\omega) = \varepsilon_0 E(\omega) + P(\omega). \quad (5)$$

Substituting (5) into (3) and writing down $\varepsilon_0 E$ as the common multiplier, we express the dielectric function of electron gas

$$\varepsilon_{33}^T(\omega) = 1 - \frac{ne^2}{\varepsilon_0 m \omega^2}. \quad (6)$$

In the absence of an external electric field $D(\omega) = 0$, therefore, it follows from (4) that an electric field inside the medium can exist under condition

$$\varepsilon_{33}^T(\omega) = 0. \quad (7)$$

Thus, at the frequency that makes dielectric function (7) vanish, electron gas may have eigenfrequencies of electrons and the electromagnetic field it is connected with.

Collective eigenfrequencies of electrons relative to the background (which neutralizes the electron charge) are called *plasma frequencies.*

Expressions (6) and (7) give the plasma eigenfrequency of an electron subsystem in a gas channel inside a current filament formed in DE

$$\omega_p^2 = \frac{ne^2}{\varepsilon_0 m}. \quad (8)$$

The concentration of electrons injected into DE is (in the SI system)

$$n = \frac{E_z \varepsilon_{33}^T \varepsilon_0}{he}. \quad (9)$$

We derive from (1) and (2)

$$\omega_p^2 = \frac{e \varepsilon_{33}^T E_z}{hm^*}, \quad (10)$$

where ε_{33}^T is the relative dielectric permeability, E_z is the averaged to DE thickness electric field strength, h is the thickness of the piezoelectric layer, m^* is the effective electron mass.

Taking $h = 10^{-6}$ m, $\varepsilon_{33}^T = 10$, $E_z = 10^4$ V/m, $m^* = 0.54m$ (the electron mass), we obtain from (10) $\omega_p = 57$ GHz, which corresponds to the experimental data for a GaAs submicron field transistor; delay time is 2 ps, delay power = 0.1 fJ [Bozler C.O. Alley G.D. The permeable base transistor and its application to logic circuit. *Proc. IEEE.* 1982. Vol.70. No.1, p. 46-52].

Dielectric degradation in solid bodies has a resonance wave character. The physical model of these processes is based on the theory of nonlinear dynamic systems [Bogomol'nyi V.M. To dynamical theory of electrothermal degradation and non-destructive testing (NDT) of defects in MDM structures. "Thermosense XXI. *Part of SPIE Conf.: "Aerosense '99".* Apr. 1999. *Proc. SPIE.* 1999. Vol. 3700. p.436-444; Bogomol'nyi V.M. Resonance calculating for electrothermal damage to metal-insulator-metal structures. *Measurement Techniques.* 2000. Vol.43. No.6. p.538-543.Transl.into Engl.].

4. Dimension electrophysical effects in PZ structures

Atoms determine the properties of chemical elements, and molecules determine the physico-chemical properties of materials. Similarly, there might be an elementary particle in the condensed phase that determines mechanical properties. There is a threshold minimal number of molecules which form the cluster of the condensed phase.

Clusters may appear in the process of evaporation and spraying on a plate, in a fluid medium or directly in a solid body. Experiments have shown that such clusters have a non-homogeneous structure (with the minimal size of 1-25 nm) [Denisyuk I.Y., Meshkov A.M. Nanostructuring is a technique for optical and semiconductor media. Optical Journal. 2001. Vol. 68. No.11, p.58].

Standard physico-mechanical characteristics cannot be realized in nano-sized solid particles. The minimal size of a nucleus obtained by laser spraying of ultradisperse diamond is ~2.5 nm. The minimal size of diamond nuclei grown from hydrogen-methane mixture on silicon is ~5 nm. The increase in size may be explained by the effect of the plate and relatively high condensation temperature (850°C). The minimal size of tin drops falling off the tip under a strong electric field is ~2 nm. Structural defects in metal alloys have a similar size. [Vesnin Y.I. Secondary Structure and Properties of Crystals. Novosibirsk: 1997(in Russian)]. According to Y.I.Vesnin, the minimal size of the order dislocation as a constant crystal lattice is physically incorrect. In the real situation the minimal size of micropores and dislocations is comparable to the size of a cluster of several hundred atoms.

Most (60-70%) atoms in nano-sized elementary crystallites (grains) are located on the surface. This feature determines the unusual properties of nano-sized devices in microelectronics and high plasticity of construction materials with ultradisperse structure (metals, ceramics).

The physics of atoms joining into mechanically stable clusters, which contain dozens and hundreds of atoms, is poorly known. It is difficult to build a common physico-mathematical model for all known experimental data (each instance reveals different sizes and shapes of atom clusters).

To estimate the size of a cluster one may use the electrochemical potential. A group of molecules (atoms) may become a solid-state cluster when the energy of molecular interaction is comparable to the electrochemical potential. The minimal size of the cluster is determined from the condition that the charge q creates a contact potential difference $\varphi = q / \varepsilon C$ on its capacity C in a medium with a dielectric permeability ε.

When a crystal cluster is formed in a gaseous medium, it is necessary to take into account the electron affinity of the cluster atoms instead of the difference of electrochemical potentials.

Based on the electron configuration model of solids, the effect of the electron subsystem on the formation of a dynamically stable cluster from dozens and hundreds of atoms is proved experimentally by the analysis of metal alloy microhardness. For example, the hardness of Ti-V-Cr alloy depends on the number of electrons in the d-shell of the atoms and increases as it grows. Thus, the electrons determine the "covalent component in the resonance covalent-chemical bond" [Samsonov G.V. et al. Configuration Model of the Substance. Kiev: Navukova Dumka. 1971 (in Russian)].

For an ideally pure crystal the size of the elementary solid particle (from the model of electron localization in a solid body) can be calculated by the radius of the spherical domain where the heat electron combines with the atom of the crystal lattice r_T

$$r_T = \frac{2e^2}{3\varepsilon_{33}^T \varepsilon_0 kT} \qquad (1)$$

where e is the electron charge, $\varepsilon_{33}^T \cdot \varepsilon_0$ is the absolute dielectric permeability, k is the Boltzmann's constant, T is the temperature (in energy units) [Bogomol'nyi V.M. Resonance calculation for electrothermal damage in metal-insulator-metal structures. Measurement Techniques. 2000. Vol.43. No.6, p.538-543].

The resonance nature of chemical bonding was first mentioned by P.N. Lebedev.

The Institute of Theoretical and Applied Mechanics of the Russian Academy of Sciences discovered a technique of "cold gas-dynamic spraying". They found three intervals for the initial velocities of a cluster (a ~10 nm solid particle), which differ by the mechanism of collision with a plate of copper and nickel clusters: the interval up to 300 m/s is the elastic rebound, 300-500 m/s denote the elastic-plastic strains, above 500 m/s denote plastic strains with nearly total dissipation of initial kinetic energy. Subsequently, clusters of aluminum were sprayed on a nickel plate, forming intermetallide Ni3Al. Diffraction studies with synchrotron radiation demonstrated the formation of a 20-50 nm Ni3Al film, which indirectly testifies to the presence of mechanically stable solid particles. The "cold spraying" technique can be used to fabricate electrodes on piezoelectric materials and in interconnections of electronic circuits. Solid piezosemi-conductors and DE contain micro- and macrodefects of various physics (impurities, dislocations, pores, cracks), which develop as a result of uncontrolled fluctuations of technological parameters in the manufacture of heterogeneous solid-state electronic devices [36].

Contemporary industry makes use of amorphous piezoelectrics, which have advantages in comparison with crystals. Their application helps to fabricate new devices [118].

If in a non-homogeneous medium the scale of non-homogeneity is comparable to such a microscopic parameter as the electron run length, we may speak about electroconductivity at a given point $\sigma(r)$, or so-called local electroconductivity.

Effective electrophysical constants cannot always be calculated by their average values, since structural non-homogeneities greatly change the laws that characterize homogeneous materials. Defects give rise to radically new wave resonance phenomena, such as vortex plasm caused by local temperature gradients [70, 88, 89].

The simplest defects are electrically active impurities and micropores. In these cases the electroconductivity of conducting materials is calculated as the electroconductivity of the "random wire net" or on the basis of the percolation theory. The effective conductivity σ_{ef} of a porous material is calculated by the formula

$$\sigma_{ef} = \sigma\left(1 - \frac{3}{2}a\right), \quad (1)$$

where a is the part of the conductor volume occupied by pores.

Under an alternating electromagnetic field σ_{ef} depends on the frequency ω and the wavelength λ.

The effective electroconductivity of a two-phase system (a composite material with a matrix with the electroconductivity σ_0 and periodically placed inclusions with the conductivity σ_1 is calculated by the following formula [Odolevsky V.I. Calculation of generalized conductivity in heterogeneous systems. Sov. Phys. – Techn. Phys. 1951. Vol.21. No.6, p. 667-677]

$$\sigma_{ef} = \sigma_0 \left(1 + \cfrac{a}{\cfrac{1-a}{3} + \cfrac{\sigma_0}{\sigma_1 - \sigma_0}} \right), \qquad (2)$$

where a is the volume part of inclusions, in the case of cavities $\sigma_1 = 0$.

Formula (2) agrees with the results obtained from the disturbance theory.

The properties of solids also depend on structural microdefects, the size of which exceeds the free run length of current carriers. The scope of edge dislocation on electron transport in semiconductor and DE materials is limited by the Debye screening radius [3].

Real crystals have one principal peculiarity. Under sufficiently high temperature the movement of free charge carriers near defects is diffusional, i.e. greatly depends on the uneven concentration of charges (near phase interfaces, dislocations, pores and rigid inclusions).

Imperfections of the crystal lattice also influence effective electroconductivity, which can be measured with high precision. The measurements and calculations in this book can be used for studying the defect structure of crystals.

Ferroelectric (FE) films are used in memory devices, sensors, cold cathodes and vacuum electronics. The effect of film thickness on the threshold value of electron emission was studied in [Park Ji. et al. Influence of thickness on the emission threshold field of Pb(Zr0,4Ti0,6)O3 (PZT) film. Jap. J. Appl. Phys. Part 2. Letters. 2002. Vol.41, No.6A, L647-L650]. The effect of thickness on uneven polarization is considered in [Ishibashi Y. et al. Thickness transitions of ferroelectricity in thin films. J. of Phys. Soc. of Japan. 2002. Vol.71, No.6, p. 1471-1474].

The effect of increasing the breakdown strength of superthin DE layers and fibers. Multilayer isolation by Academician A.F.Ioffe

The plasticizing effect of the impulse electric current in the treatment of metals is well known. It has been established that Joule heating is not the only reason for plastic deformation. It is necessary to take into account local heating near microdefects. The electroplastic effect may be caused by dislocation travel and interaction. The electric current increases the speed of moving dislocations, helping them to overcome obstacles in the slide planes. It has been found that the deformation caused by the impulse of the electric current is greater than the deformation caused by Joule heating.

Under an external electric field metals and ion DE crystals develop a plastic deformation caused by a "forest" of dislocations. As applied to metals, this phenomenon was called the effect of "electric superplasticity". It is used for the deep pressing of thin-film parts. Electric impulses, plasmochemical spark treatment and HF induction heating increase the cyclic strength of metallic parts by 2-3 times and reduce surface roughness by 2 classes.

Under an electric field moving dislocations form dot defects (and their clusters). This changes the electron and ion conductivity of DE. For example, after 10% plastic deformation of an ion crystal the concentration of Schottky's defects reaches 1014 cm–3. The generation of dot defects is most intensive under stresses greater than the yield stress (when dislocations occur).

Under the critical values of the electric field strength (determined in § 3.1, 3.8) in MDM structures near the tops of microcracks (and rigid inclusions, the micropeaks on the surface of electrodes) we observe local plastic deformation, which redistributes internal mechanical stresses and local electric field strength. These changes, as well as and low-voltage prepolarisation of DE, increase electric strength.

Electric strengthening in solid DE allows to study electronic processes in thin DE layers without destroying them by breakdown.

A.F.Ioffe supposed that electric breakdown in glasses is due to the shock ionisation and subsequent collision of free ions. However, experimental measurement showed that the breakdown time is much smaller than the ion run time. The electronic mechanism of shock ionization was experimentally established later.

First attempts to give experimental proof to electric strengthening in thin layers of mica and glass were unsuccessful. It was necessary to fabricate thinner layers of homogeneous DE.

The effect of electric strengthening was proved in monocrystallic layers of polymer DE films with a thickness of tens of micron.   The essence of this effect is that thin layers of solid DE can tolerate (without mechanical destruction) electric fields by one or two orders exceeding the fields that usually destroy even much thicker samples.

In "superstrong" fields free electrons (in the conductivity zone) accelerate to high energies and cause shock ionization and excitation of the ions in the crystal lattice. Intensive excitation should be accompanied by reverse transfers, which may occur along with photon and phonon radiation. In critical situations one may also expect local solid-state plasma phenomena. The interaction of "hot electrons" with the crystal lattice changes the mobility of free charge carriers. It is accompanied by N- and S-shape CVC, the electrothermal and thermoelectric effects. The Gunn's effect and electron tunneling are also used in industry [Vershinin Y.N., Zotov Y.A. Overheating instability in crystal insulators under extreme electric fields.

Sov. Phys. Solid State. 1975. Vol.17. No.12, p. 3487-3494]. The analysis of CVC allows one to determine the parameters of strengthening technologies.

The most probable reasons for electric strengthening of thin-film DE are:

At the critical values of the electric field strength, determined in § 3.8, there is an additional polarization of DE. Firstly, it increases the elasticity modulus and electromechanical strength due to the change in the orientation of molecules and domains. Secondly, the concentration of electric fields near the micropeaks on the surface of metallic electrodes brings about a local increase in temperature (up to the melting point) and healing of microdefects (the tips of cracks are sealed and the microtip curvature increases).

The effect of strengthening at local temperature growth (due to electron injection into DE) at currents limited by a space charge, may be caused by the transition from the brittle destruction mechanism to the elasto-plastic or elasto-viscous one [28].

The zone of electrode influence is comparable to the electron run length in DE (for polymers 120-130 μm, for melted quartz 150-180 μm with an electron energy of 50-80 keV). Near the electrode the structure and properties of DE are very different from the volume ones. Due to the concentration of the electric field strength and sharp temperature gradient at the metal-DE contact, there are termoelastic shear and normal separation stresses. At the boundary of the electrode zone DE is covered with cracks, perpendicular to the vector of the electric field strength.

In thin layers mechanical strength is higher due to the following reasons:

In the manufacture of thin films and fibers there are by three or four orders fewer residual defects.

The influence of strong local electric fields near structural defects is similar to the effect of small doses of ionizing radiation. It also increases the strength of solid-state electronic devices.

DE destruction is a dynamic system of interconnected physical and chemical phenomena. The electrothermal breakdown of DE has a resonance wave character. Under certain conditions (see § 3.8) at the initial reversible stage of electric breakdown defects may "heal". Self-organization is a fundamental property of various nonlinear dynamic systems [94, 102, 105].

At overheating instability the electron-phonon system self-adjusts to the outer "load". It automatically rearranges electric, heat and mechanical fields and, accordingly, the structure of the material. The destruction is also characterized by an unstable dynamic balance between two opposite processes, the formation and accumulation of defects on one hand and their "healing" on the other.

The theory of nonlinear dynamic systems can help to explain the electric strengthening of thin DE layers in the following way:

There is local polarization near micropeaks on the electrode surface in DE in the non-polar phase, and additional polarization in active DE. Hence, molecules in the external constant field undergo additional re-orientation (which is what increases the strength). Due to this MDM structures become unipolar, which creates autooscillations under an external constant field. At the initial reversible stage of electric breakdown the autooscillations of the electromagnetic field and current can "heal" of structural defects. The techniques of eliminating technological defects in impulse, HW and USW fields are used in industry. In thin-layer MDM structures one observes (in relatively weak fields) thermoelectronic and autoelectronic emission from micropeaks on the surface of metal elec-trodes. The DE diode operates in the mode of injection currents, limited by the space charge.

In a high ohmical DE medium near micropeaks on the surface of metal electrodes there are "current filaments" (channels of electroconductivity) and DE Joule heating. This in its turn increases the density of the injection current. This interconnected ("circular") process of "swinging" the electron subsystem abruptly increases the electron energy and local temperature near "current filaments". Current and temperature positive feedbacks result in ferroelectric instability of the electron-phonon system and S-shape CVC.

A flow of electrons injected into DE, compensated by the space charge of immobile positive ions in the DE crystal lattice (bound charges) and free ions (which appear as temperature rise), drifting towards the cathode, forms quasineutral low-temperature plasma. As follows from experimental data [33, 47], random fluctuations connected with structural defects in-crease because of "overheating" electric instabilty. Near the fist cusp of the S-shape CVC the electric current oscillates, which in certain cases is ac-companied by electromagnetic wave radiation from DE into the environ-ment in a wide frequency range. The unipolarity effect may be a reason for autooscillations [Golovin Y.I. Characterization of plastic deformation and ion crystal destruction by eigen electromagnetic radiation. Condensed Me-dia and Interfaces. 2002. Vol.4. No.1, p.5-16]. Current oscillations and sharp temperature increase determine the onset of the irreversible stage of the electrothermal breakdown in a DE, when micropores, dislocations and microcracks appear.

At harmonic oscillations in some localities of a polar DE, apart from Joule heating, near "current filaments" and in DE bridges between pores and cracks (near the tops of cracks) the heat reaches the temperature of phase transition. This happens due to energy dissipation caused by dielectric, mechanical and piezoelectric losses (see § 3.4). At the same time, the tips of the cracks are sealed, which increases the curvature radius of crack tops and correspondingly decreases the concentration of electric and mechanical fields near the crack tops.

A local temperature growth automatically leads to the reconstruction of the surface of metal electrodes. The tops of microtips melt, the curvature radius increases. As shown in § 3.8, this process induces the self-healing of structural defects under an external electric field smaller than the critical value.

Critical values of the electric field strength determine the bifurcation point (which determines whether destruction is reversible).

Academician A.F.Ioffe proposed multi-layer high-voltage insulation with alternated DE and conductor layers (metal foil). The technology of multi-layer insulation was surprisingly simple: the foundation was gradually covered with drying layers of natural oil. In air the thin surface layer oxidized and acquired dielectric properties. The inner layer maintained its electroconductivity

A.F.Ioffe supposed that electric breakdown is caused by avalanche ionization in crystals. In his view, in glasses (and calcites) sodium ion accelerates in the external field, and a non-elastic collision with the ion (or atom) of the crystal lattice forms an ion pair. Ten collisions give 1024 new ions (due to a chain reaction). A.F.Ioffe suggested making the DE thickness smaller than the ion run length so as to avoid the chain reaction of ion multiplication.

From later experiments it became clear that most insulators have an electron avalanche, and the DE thickness has to be smaller that the electron run length. For melted quartz the path of electrons with an average energy of 50 keV is approximately ~120 μm. The thickness of oil layers in A.F.Ioffe's experiments was comparable to this value.

The effect of strengthening multi-layer insulation may be due to Maxwell-Wagner's interlaminar polarization (which occurs under a constant electric field near the interface of materials with different conductivities). The molecules in this case are additionally oriented towards the vector of the electric field strength (along the DE thickness). According to experimental data, polarization increases by one order the elasticity modulus of a polymer matrix with fiberglass [Lusheikin G.A. Modelling of elastic properties of glass-filled composites. Plastic masses. 2001. No.5, p.17 (in Russian); Yemets Y.P. Electric forces on the interface of DE media. Journal of Applied Mechanics and Technical Physics. 1993. Vol. 34. No.4(200), p. 14].

Metallic layers in A.F.Ioffe's multi-layer insulation screen the external electric field (a similar effect may occur due to interlaminar polarization).

Molecular layers, Langmuir-Blodget films, have an anomalously high strength and are used in the last generation of superfast computing devices, as well as chemical sensors.

Filamentary crystals of silicon nitride (SN) with a diamond cube lattice are a unique material for transforming changes of temperature, pressure and accelerations into an electric signal (tensometers, accelerometers, thermistors). Tension causes filamentary crystal torsions (due to accumulated screw dislocations). Measuring the torsion angle allows us to increase the sensor sensitivity by 4-5 times compared to measuring the tension strain (see the SN strength Table). The strain at SN destruction is 4.5%, which can be compared to the permissible elasto-viscous deformation of capron filaments.

Type of load	SN sample diameter $d \cdot 10^6$ m	Strain at failure ε_{max}, %	Strongest SN diameter $d \cdot 10^6$, m	Reference value ε_{max}, %	Ultimate stress limit at 300 K $\sigma \cdot 10^{-9}$
Tension	$1 \div 80$	4.5	1	3.6	8.3
Torsion	$4 \div 45$	2.9	4	2.8	2.3
Bending at distributed load	$2 \div 68$	3.5	2	2.5	6.5
Two-point bending	$10 \div 108$	2.5	10	–	4.6

Table 7. SN strength at 300 K

Semiconductor crystals exhibit an unusual increase in conductivity at compression (up to 109 times). Electroconductivity grows noticeably at mechanical stresses above the yield stress. The transport of free charges is localized near the slipbands.

In 1933 A.V.Stepanov discovered the charge appearance on the surface of rock salt crystal due to plastic deformation. This creates surplus charges of the opposite sign on the "shores" of appearing cracks and leads to electric discharges, electromagnetic radiation (EMR) ranging from radiowaves to visible light, emission of high-energy electrons (hundreds of keV) and other types of radiation (X-ray, gamma ray). The radiation activity of newly formed surfaces causes chemical activity (which determines the aging of materials).

Electromagnetic radiation (EMR) occurs at brittle destruction of a crystal due to acoustic oscillations of crack shores (c.f. electrode oscillation in an air condenser). Since the radiation frequency equals the absorption frequency, measuring EMR frequencies at DE failure by radiotechnical devices can help to choose the operating frequency for healing structural defects and stopping the developing cracks by electromagnetic fields.

Porous silicon (PS) is a new and promising material. It is used for insulating localities in integrated circuits. The discovery of visible intensive luminescence of PS led to the creation of optoelectronic devices and measuring devices. The resistance of PS with boron admixture at thermal exposure (500°C) abruptly drops (by 4-5 orders), which may be used in temperature sensors [Properties of porous silicon. Ed. L. Canham. EMIS. Data Reviews. 1997, No.18; Appl. Phys. Lett. 1990, v. 57, № 10, p. 1046-1048]. PS electroconductivity can be calculated by formulas (1) and (2).

5. Synchronization in autooscillating systems with DE resonators

Radiotechnical systems of radiolocation, satellite communication, frequency and time standards contain highly stable sources of oscillations with a pure wide-range spectrum. It creates the need for precision measuring devices, such as signal generators, spectrum analyzers, frequency and amplitude fluctuation gauges.

The most common way to decrease frequency noise in generators is to increase the quality of the oscillating system. There is a technique of stabilization based on the effect of frequency capture by a high-quality resonator. The frequency capture effect is characterized by the stabilization coefficient S, which is calculated as the ratio of the retuning of the partial frequency of the generator contour to the frequency change caused by this retuning. The simplest and at the same time optimal is the construction of three-circuit MW generators with monocrystal leucosapphire dielectric disc resonators (DDR) with a quality $Q = 2.6 \cdot 10 5$ at $T = 300$ K.

However, the practical application of DDR as standards for MW generators is limited due to their high frequency temperature coefficient (FTC). Thus, to maintain a constant resonance frequency, we need to thermostate DDR or synchronously retune it opposite to frequency drift. It is more practical to tune the DDR frequency by temperature control.

Some instances require powerful sources of coherent radiation with a phase difference that can be accurately fixed. Autosynchronization, when the powers of several generators are combined in the common load, helps to increase the power of MW generators. The analysis of noise in joined generator systems showed that autosynchronization is realized at slow frequency fluctuations.

Using leucosapphire DDR simultaneously as a resonance connection element and stabilizing resonator can help to create a set of highly stable three-contour autogenerators (AG).

The DDR frequency can be changed in two main ways: by changing temperature (due to high FTC), as well as with the effect of metal and DE surfaces on the external field of resonators [Vzyatyshev V.F., Kalinichev V.I., Kuimov V.I. Characteristics of the planar metal-dielectric resonator. J. Communs. Technology Electronics. 1985. No.8, p.1549-1553; Dobromyslov V.S. Oscillations in a metal dielectric resonator with a leucosapphire bar. J. Communs. Technology Electronics. 1988. Vol.33. No.4, p.705-716; Chyorny B.S., Ilchenko M.E. Effect of metal surface on the properties of an open MW dielectric resonator. Sov. Radioelectronics Journal. 1978. Vol.21. No.8, p.52-59].

To decrease the control "inertia" of synchronization parameters, one can use a heat-radiating metal mirror placed under DDR. Its temperature is created by a thermal element due to the Peltier effect (see Fig. 31).

A much simpler solution is the construction of automatic frequency tuning (AFT) in which changes of DDR temperature are determined by measuring the frequency differences between different DDR oscillation modes.

Automatic frequency tuning (AFT) based on the effect of influencing the electromagnetic field distanced from dielectric disc resonators (DDR).

The simplest and most precise method to determine the resonance frequency of DDR is to use it as an external resonator in "frequency capture" circuits. A layout of AFT with a three-contour generator based on a Gunn's diode, stabilized by a leucosapphire DDR (diameter D = 67 mm, h = 4 mm, fr = 10 GHz, stabilization coefficient S = 500) is given in Fig. 31. The generation frequency through the frequency transformer is sent to the frequency meter (which measures frequency with τmeas = 1 s). The information about frequency was sent from the outlet of the frequency meter in the binary decimal code through two channels to the electronic block of autoregulation. The dynamic range of the frequency control system is determined by the digit capacity of a digital-to-analog converter (DAC). For a 12-digit DAC the range amounts to 4096 conditional units. To increase the range, the indication of the frequency meter was registered from the second digit. At averaging time τmeas = 1 s (the last digit was 1 Hz) the dynamic range was 40 kHz.

In the standard AFT circuit one usually eliminates the consequence rather than the cause. At a steady DDR frequency drift the system leaves the measurable range. To eliminate this disadvantage, the frequency in the AFT is artificially "swung" with the help of a heater (wound like a spiral wire onto the metallic element of the DDR mount). When the electron circuit impulse counter was overfilled, the heater was sent a voltage impulse through the emitter repeater. As soon as the frequency returned into the regulation band, the heater was switched off, i.e. the experiment used an ordinary relay system for thermal regulation.

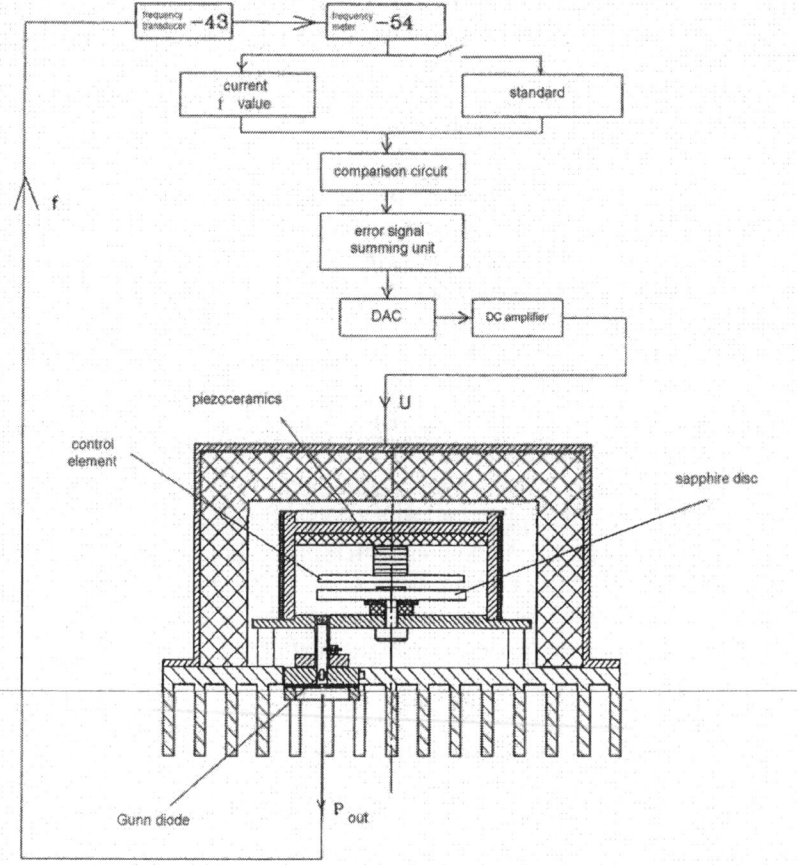

Fig. 31. Layout of AFT based on the influence on the field distanced from DDR

After the generator entered the stationary operation mode by heat (frequency drift Δf < 100 Hz per 1 s), the electronic comparison circuit disconnected the circuit of one channel, and the last value was memorized. This value later became the frequency "standard".

The error signal from the DAC outlet, proportionate to the difference between the current and standard frequency, was sent to the electrodes of a piezoceramic element (a set of piezoceramic thin discs) through the block of ftemperature rise control. At voltage change from 0 to 300 V and piezo-element thickness H = 45 mm a thin copper disc (on the face of the piezo-element) was moved to the maximal value $\Delta Uz \approx 1.2$ µm. The working point of voltage on the piezoelement was chosen on the linear section of dependency $\Delta Uz(U)$ at U = 150 V, which corresponded to the middle of the DAC scale. The AFT came to registering the position of the metal mirror at frequency changes ≈ 40 kHz, when the supply voltage changed in the range of 300 V. The use of piezoceramics enables one to modulate the DDR frequency by a given control function to several kHz.

Synchronization devices are the main block of receiving, transmitting and measuring radiotechnical and electrical systems. They determine their precision and noise resistance.

The simplest mathematical model of autooscillation systems (AS) contains an algebraic nonlinearity, which is studied by creating a planar phase portrait of a nonlinear dynamic system. The dynamics of processes in phase frequency autotuning devices is described by a model which contains one nonlinear periodic function (determined by the characteristic of a phase discriminator) and is analyzed by building a phase portrait on the cylindrical phase surface, i.e. cylindrical phase systems [Thompson J.M.T., Virgin L.N. Predicting a jump to resonance using transient maps and beats. Int. J. Nonlinear Mechanics. 1986. Vol.21. No.3, p.205-216].

In synchronization devices direct capture helps to achieve stabilization, frequency division (multiplication) and signal filtration. Synchronization based on phase frequency autotuning differs from direct capture. It is determined by the presence of electronic circuits, i.e. the block of error signal formation and the block of frequency control. This allows us, apart from the functions mentioned, to perform angle modulation and obtain a net of stable frequencies (frequency synthesis) if the "ring" is completed with a variable coefficient divider. The effect of accidental fluctuations on the synchronization of oscillations in dynamic systems is considered elsewhere.

The method of phase synchronization of polar DE-based resonators, given in § 3.5, allows one to simplify phase frequency autotuning due to the understanding of the physics of this phenomenon.

6. Posistor ceramics in automatic control systems

The history of electronic ceramics began with the discovery of ferroelectric, barium titanate. This material radically changed our views on properties, application and industrial technology of solid-state electronic devices. The dielectric permeability of barium titanate is 10-100 times greater than that of rutile. This challenged the idea of maximally achievable capacity of condensers.

After introducing a microscopic amount of admixtures into barium titanate ceramics, its electric resistance is reduced by 8 orders, and DE becomes a semiconductor (posistor ceramics). In common semiconductors electric resistance decreases as temperature rises; e.g., in barium titanate at 120°C (which corresponds to phase transition) resistance, on the contrary, increases by 103 and in some cases by 107 times. This is used in constant-temperature heaters (hair dryers, oil and gas heaters in automobile engines and thermoregulators).

When barium titanate-based ceramics is baked, it acquires various electroconductivity, not only due to admixtures, but also depending on the material purity and technology parameters.

The specific resistance of barium titanate (used as DE or piezoelectric) at normal temperature is ~108 Ohm·m. After adding rare-earth elements, its specific resistance reaches 0.1-104 Ohm·m (and coincides with the resistance of semiconductors). A semiconductor barium titanate with an indium or gallium electrode (which forms an ohmical contact) and a silver electrode becomes a rectifying diode. It gives a CVC which allows one to use a posistor as a voltage or current stabilizer.

Semiconductor barium titanate has the following properties [1, 9, 11, 12]:
- its electric resistance largely nonlinearly depends on electric voltage (the varistor effect),
- it forms with common electrodes a large barrier capacity (thus may be used in condensers).

A posistor has two functions, a heater and a temperature regulator. Hence, it is used as a thermostabilizer which is not "afraid" of overheating (electric cookers, fireplaces, Thermoses).

Posistors are used to limit the current, to test the level of liquids, to start electromotors, for automatic demagnetization in colour televisions, in time relays and signal delay elements [5].

Russian starter relays in refrigerators are 20 mm-ceramic discs with a thickness h = 2.5 mm and the following parameters: Rel = 3.3; 22; 27; 47 Ohm, Uop = 115; 220 V, Istart = 8.14 A. Fuel heaters for automobiles have a diameter of 16 mm, h = 2 mm and the following parameters: Rel = 3.3±20% Ohm, Top = 130±10°C, Uop = 12 V. Russian posistors have the following characteristics:

- switch temperature 50-220°C,
- nominal resistance $(1 \div 2) \cdot 104$ Ohm,
- threshold electric resistance up to PTRC 103-105,
- maximal nominal operating voltage 380 V,
- maximal electric field strength 200 V/mm (for PZT-4 piezoceramics the maximal electric field strength is 100 V/mm).

Below the Curie point the potential barrier on the interface becomes lower due to spontaneous polarization. Above the Curie point there is no polarization, the potential barrier grows and resistance increases.

The introduction of powdered MnO2 and CuO into ceramics gave high dielectric permeability, which showed the presence of a polar DE with a barrier layer. Consequently, the piezoelectric effect is possible only in posistor ceramics.

7. Electrophysical properties of solid interfaces

The development of microelectronics and other fields of technology (heterogeneous catalysis, material strength) calls for the study of electronic properties near interfaces of heterogeneous solids. These issues are relatively unknown (e.g., the calculation of the bond energy of atoms and ions adsorbed on the surface of metals, the adhesion theory of metals and DE and semiconductors) [120].

A rigorous mathematical description of these interface problems is limited by the absence of translation symmetry and the necessity for taking into account the electron-phonon interaction. This does not allow for traditional methods, which were designed for the analysis of volume (integral) properties of solids [145-147].

The formulated physical model gives an idea of how electronic states influence chemical bonds in clusters consisting of two or three thousand atoms at alloying and under ionizing radiations [Gubanov A.I., Dunaevsky S.M. Non-metal and ion crystal adhesion. Sov. Phys. Solid State. 1977. Vol.29, p.1369-1376; Gubanov A.I., Dunaevsky S.M. Adhesion of two ion crystals. Sov. Phys. Solid State. 1976. Vol.18, p.2248-2256]. For example, the adhesion energy of the composite Al-NaCl was found to be $E_a = 0.77$ J/m2. The strongest adhesion bond occurs between metal with a higher density of free electrons and ion crystal.

Electrochemical potential is a fundamental parameter of any material. A group of atoms (molecules) may be a cluster of a new solid phase at contact difference of potentials between two particles. The minimal size of the solid-state nucleus is determined from the condition that a singular charge q on its capacity C with a relative dielectric permeability ε_{dp} creates a contact difference of potentials. When a solid particle is formed in the gaseous phase, one should take into account the energy of electron affinity to the atoms of the particle "embryo" rather than the difference of electrochemical potentials.

8. Construction and calculation of piezotransducers

A resonance sensor of detonation is fabricated according to the layout in Fig. 48, with bending strains of the bimorph element.

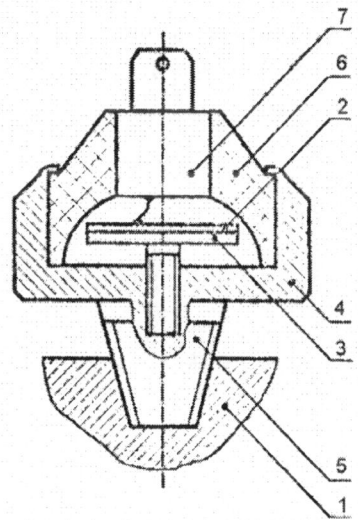

Fig. 48. Construction of a detonation sensor (1 is the engine fragment; 2 is the piezoelement; 3 is the elastic membrane; 4 is the shell; 5 is the threaded fastening bolt; 6 is the lid; 7 is the connector)

The sensor of detonation in internal combustion engine 1 is mounted in the cylinder shell 1 with the fastening bolt 5, and part of its threading is the receiving link. The other part of bolt 5, shell 4 and elastic membrane 3 with their own bolt are a transmission link of the input influence, whereas the piezoelement 2 and elastic membrane 3 are the sensitive element. The correct reproduction of the measured parameter in time $U(\tau)$ largely depends on the characteristics of the sensitive element and its operation mode. At the same time, both shell 4 and bolt 5, being elements of other links in the structure, can distort the transmitted acoustical signal. In the common case the input circuits of the electron block in the engine control system switch on the cable line with the capacity C_k and inductivity L, as well as input electric resistance R_n and capacity C_n, connected parallel to the cable. The construction of detonators contains a shunting resistor (3-10 kOhm).

Oscillation equations of piezoceramic elements (PCE) in cylindrical coordinates

Using cylindrical coordinates, we align the axis x_3 to the direction of the piezoceramics polarization vector and use cylindrical coordinates r, θ and z: $x_1 = r \cdot \cos(\theta)$, $x_2 = r \cdot \sin(\theta)$, $z = x_3$. The components of the mechanical strains tensor, expressed in cylindrical coordinates, are connected with tensor components in cylindrical coordinates [75]

$$\sigma_{rr} = \cos^2(\theta) \cdot \sigma_{11} + 2 \cdot \sin(\theta) \cdot \cos(\theta) \cdot \sigma_{12} + \sin^2(\theta) \cdot \sigma_{22},$$

$$\sigma_{\theta\theta} = \sin^2(\theta) \cdot \sigma_{11} - 2 \cdot \sin(\theta) \cdot \cos(\theta) \cdot \sigma_{12} + \cos^2(\theta) \cdot \sigma_{22},$$

$$\sigma_{r\theta} = \sin(\theta) \cdot \cos(\theta) \cdot (\sigma_{22} - \sigma_{11}) + (\cos^2(\theta) - \sin^2(\theta)) \cdot \sigma_{12}, \qquad (1)$$

$$\sigma_{rz} = \cos(\theta) \cdot \sigma_{13} + \sin(\theta) \cdot \sigma_{23},$$

$$\sigma_{\theta z} = -\sin(\theta) \cdot \sigma_{13} + \cos(\theta) \cdot \sigma_{23},$$

$$\sigma_{zz} = \sigma_{33}.$$

In the common case, at axis polarization (directed towards z) the equations of electroelasticity have the form:

$$\sigma_{rr} = c_{11}^E \cdot S_{rr} + c_{12}^E \cdot S_{\theta\theta} + c_{13}^E \cdot S_{zz} - e_{31} \cdot E_z,$$

$$\sigma_{\theta\theta} = c_{12}^E \cdot S_{rr} + c_{11}^E \cdot S_{\theta\theta} + c_{13}^E \cdot S_{zz} - e_{31} \cdot E_z,$$

$$\sigma_{zz} = c_{13}^E \cdot (S_{rr} + S_{\theta\theta}) + c_{33}^E \cdot S_{zz} - e_{33} \cdot E_z,$$

$$\sigma_{z\theta} = 2 \cdot c_{44}^E \cdot S_{z\theta} - e_{15} \cdot E_\theta,$$

$$\sigma_{rz} = 2 \cdot c_{44}^E \cdot S_{rz} - e_{15} \cdot E_r,$$

$$\sigma_{r\theta} = (c_{11}^E - c_{12}^E) \cdot S_{r\theta}, \qquad (2)$$

$$D_r = \varepsilon_{11}^S \cdot E_r + 2 \cdot e_{15} \cdot S_{rz},$$

$$D_\theta = \varepsilon_{11}^S \cdot E_\theta + 2 \cdot e_{15} \cdot S_{\theta z},$$

$$D_z = \varepsilon_{33}^S \cdot E_z + e_{31} \cdot (S_{rr} + S_{\theta\theta}) + e_{33} \cdot S_{zz},$$

$$S_{rr} = \frac{\partial u_r}{\partial r}, S_{\theta\theta} = \frac{1}{r} \cdot \frac{\partial u_\theta}{\partial \theta} + \frac{u_r}{r}, S_{zz} = \frac{\partial u_z}{\partial z},$$

$$S_{r\theta} = \frac{1}{2} \cdot \left(\frac{1}{r} \cdot \frac{\partial u_r}{\partial \theta} + \frac{\partial u_\theta}{\partial r} - \frac{u_\theta}{r} \right), \ S_{rz} = \frac{1}{2} \cdot \left(\frac{\partial u_z}{\partial r} + \frac{\partial u_r}{\partial z} \right),$$

$$S_{\theta c} = \frac{1}{2} \cdot \left(\frac{\partial u_\theta}{\partial z} + \frac{1}{r} \cdot \frac{\partial u_z}{\partial \theta} \right),$$

$$\mathbf{E_r} = -\frac{\partial \mathbf{U}}{\partial \mathbf{r}}, \ \mathbf{E_\theta} = \frac{1}{\mathbf{r}} \cdot \frac{\partial \mathbf{U}}{\partial \theta}, \ \mathbf{E_z} = -\frac{\partial \mathbf{U}}{\partial \mathbf{z}}.$$

When a flat longitudinal wave normally falls on PCE as a disc with solid electrodes on its end faces, some components of the tensors of strains and electric field strength are zero:

$$\frac{\partial \mathbf{u_z}}{\partial \mathbf{r}} = 0, \ \frac{\partial \mathbf{u_r}}{\partial \theta} = 0, \ \frac{\partial \mathbf{u_\theta}}{\partial \theta} = 0, \ \frac{\partial \mathbf{u_z}}{\partial \theta} = 0, \ \frac{\partial \mathbf{u_r}}{\partial \mathbf{z}} = 0, \ \frac{\partial \mathbf{u_\theta}}{\partial \mathbf{z}} = 0, \ E_r = 0,$$

$$E_\theta = 0.$$

The equations of electroelasticity in the volume stressed state of such a PCE take the form:

$$\sigma_{rr} = c_{11}^E \cdot \frac{\partial u_r}{\partial r} + c_{12}^E \cdot \frac{u_r}{r} + c_{13}^E \cdot \frac{\partial u_z}{\partial z} - e_{31} \cdot E_z,$$

$$\sigma_{\theta\theta} = c_{12}^E \cdot \frac{\partial u_r}{\partial r} + c_{11}^E \cdot \frac{u_r}{r} + c_{13}^E \cdot \frac{\partial u_z}{\partial z} - e_{31} \cdot E_z,$$

$$\sigma_{zz} = c_{13}^E \cdot \left(\frac{\partial u_r}{\partial r} + \frac{u_r}{r} \right) + c_{33}^E \cdot \frac{\partial u_z}{\partial z} - e_{33} \cdot E_z,$$

$$\sigma_{z\theta} = 0, \sigma_{rz} = 0, \qquad\qquad\qquad (3)$$

$$\sigma_{r\theta} = (c_{11}^E - c_{12}^E) \cdot \left(\frac{\partial u_\theta}{\partial r} - \frac{u_\theta}{r} \right),$$

$$D_r = 0, D_\theta = 0,$$

$$D_z = \varepsilon_{33}^S \cdot E_z + e_{31} \cdot \left(\frac{\partial u_r}{\partial r} \right) + e_{33} \cdot \frac{\partial u_z}{\partial z}.$$

Movement equations in cylindrical coordinates have the form:

$$\rho_p \cdot \frac{\partial^2 u_r}{\partial t^2} = \frac{\partial \sigma_{rr}}{\partial r} + \frac{1}{r} \cdot \frac{\partial \sigma_{r\theta}}{\partial \theta} + \frac{\partial \sigma_{rz}}{\partial z} + \frac{(\sigma_{rr} - \sigma_{\theta\theta})}{r},$$

$$\rho_p \cdot \frac{\partial^2 u_\theta}{\partial t^2} = \frac{\partial \sigma_{r\theta}}{\partial r} + \frac{1}{r} \cdot \frac{\partial \sigma_{\theta\theta}}{\partial \theta} + \frac{\partial \sigma_{\theta z}}{\partial z} + \frac{2 \cdot \sigma_{r\theta}}{r}, \qquad (4)$$

$$\rho_p \cdot \frac{\partial^2 u_z}{\partial t^2} = \frac{\partial \sigma_{rz}}{\partial r} + \frac{1}{r} \cdot \frac{\partial \sigma_{\theta z}}{\partial \theta} + \frac{\partial \sigma_{zz}}{\partial z} + \frac{\sigma_{rz}}{r}.$$

Equation systems (2), or (3), or (4) offer enough relations to study the stress-strain state of PCE in the shape of cylinders polarized towards the axis z.

At radial polarization (in the direction of r):

$$\sigma_{rr} = c_{13}^E \cdot (S_{zz} + S_{\theta\theta}) + c_{33}^E \cdot S_{rr} - e_{33} \cdot E_r,$$

$$\sigma_{\theta\theta} = c_{13}^E \cdot S_{rr} + c_{11}^E \cdot S_{\theta\theta} + c_{12}^E \cdot S_{zz} - e_{31} \cdot E_r,$$

$$\sigma_{zz} = c_{13}^E \cdot S_{rr} + c_{12}^E \cdot S_{\theta\theta} + c_{11}^E \cdot S_{zz} - e_{31} \cdot E_r,$$

$$\sigma_{z\theta} = (c_{11}^E - c_{12}^E) \cdot S_{z\theta},$$

$$\sigma_{rz} = 2 \cdot c_{44}^E \cdot S_{rz} - e_{15} \cdot E_z, \qquad\qquad (5)$$

$$\sigma_{r\theta} = 2 \cdot c_{44}^E \cdot S_{r\theta} - e_{15} \cdot E_\theta,$$

$$D_r = \varepsilon_{33}^S \cdot E_r + e_{31} \cdot (S_{\theta\theta} + S_{zz}) + e_{33} \cdot S_{rr},$$

$$D_\theta = \varepsilon_{11}^S \cdot E_\theta + 2 \cdot e_{15} \cdot S_{r\theta},$$

$$D_z = \varepsilon_{11}^S \cdot E_z + 2 \cdot e_{15} \cdot S_{rz}.$$

Often, considering PCE thin discs or rings loaded at end faces, research-ers tend to ignore the radial and circumferential oscillation modes ($u_r = 0$, $u_\theta = 0$) and suppose that all values depend on one coordinate z. This ap-proach is justifiable when the PE thickness is considerably smaller than the wavelength of the acoustic signal. It should be noted, however, that in this case the sought electric strength at the input of the electronic control block U_n or the coefficient of transforming amplification into voltage ξ_U will be slightly excessive (up to 2-6 dV), since the electric constants e_{33} and e_{31} have different signs and in the description of the volume stressed state of PE the corresponding items $e_{31} \cdot s_r$ and $e_{31} \cdot s_\theta$ decrease the electric in-duction D_z. When calculating the stress-strain state of PCE, one can omit the indices of the tensor components. The state equation in scalar terms will have the form

$$\sigma = c_{33}^E \cdot S - e_{33} \cdot E$$

$$D = e_{33} \cdot S + \varepsilon_{33}^S \cdot E$$

The wave equation

$$\rho_p \cdot \frac{\partial^2 u}{\partial t^2} = c_{33}^D \cdot \frac{\partial^2 u}{\partial z^2}.$$

At uneven PCE strains, one should compile a system of boundary equa-tions. In a specific case, when considering fewer independent components of mechanical stress and electric induction tensors, one can simplify the equation systems (1, 2 and 5).

Calculation of piezoelectric plates for wavelengths comparable to cross-section sizes

There are hypotheses for circular thin elastic plates with an axisymmet-ric bending load. The bending strains in the middle surface are negligibly small; the cross-sections that are perpendicular to the middle surface re-main flat at bending (Kirchhoff-Love hypotheses)

$$\sigma_{zz} \approx 0, \; S_{rr} = -z \cdot \frac{\partial^2 u_z}{\partial r^2}, \; S_{\theta\theta} = -\frac{z}{r} \cdot \frac{\partial u_z}{\partial r}.$$

The equation of the free oscillations of the plate has the form

$$\rho \cdot \frac{\partial^2 U}{\partial t^2} + \frac{h^2 \cdot E}{12 \cdot (1 - \mu^2)} \cdot \Delta^2 U = 0, \qquad (6)$$

where ρ is the plate density; U are the normal displacements of the neutral plate surface; h is the plate thickness; E is the equivalent elastic-ity modulus of the composite plate; μ is the equivalent Poisson's coefficient; Δ is the two-dimensional Laplace operator. For axisymmetric oscillations of the plate the operator Δ is written as

$$\Delta = \frac{1}{r}\frac{\partial}{\partial r}\left(r\frac{\partial}{\partial r}\right).$$

The calculation is reduced to finding the dependence of its steady oscillations' amplitude U on radius r at a given frequency.

The edge conditions correspond to the case when the plate is rigidly mounted on a support bar or in the shell, which oscillate in the normal direction at a frequency ω. The solution of equation (6) has the form

$$U(t,r) = U(r)\cdot\exp(-i\omega t). \quad (7)$$

Substituting (7) into equation (6) and using dimensionless variables, we obtain the equation

$$\frac{d^4u}{dx^4} + \frac{2}{x}\cdot\frac{d^3u}{dx^3} - \frac{1}{x^2}\cdot\frac{d^2u}{dx^2} + \frac{1}{x^3}\cdot\frac{du}{dx} = \alpha\cdot u, \quad (8)$$

where $\alpha = \dfrac{12\cdot(1-\mu^2)\cdot\rho\cdot\omega^2\cdot R^4}{E\cdot h^2}$, $x = r/R$, $u = U/a$.

Within $\omega \to 0$ the dimensionless coefficient $\alpha \to 0$. Equation (8) is reduced to the known equation for the static deformation of a circular plate and has the following solution:

$$u_0(x) = B_1 + B_2\cdot x^2 + B_3\cdot\ln(x) + B_4\cdot x^2\cdot\ln(x). \quad (9)$$

The constants $B_1 - B_4$ are determined by the edge conditions.

The calculation of a piezotransducer at bending strain requires given values of mechanical stresses, forces or moments at the corresponding boundaries of the plate (see Fig. 49). Thus, e.g., for a piezoceramic plate mounted on a central bar, on the external boundary of the plate R_0 the bending moment in the radial direction M_r and the shear force Q_r are zero. On the internal contour r_0 one should put to zero the displacement u_z and deflection angle $\psi = \partial u_z / \partial r$ (rigid fixed boundary, see Fig.48)

$$\frac{\partial^2 u_z}{\partial r^2} + \frac{\mu}{r}\cdot\frac{\partial u_z}{\partial r} = 0|_{r=R_0}$$

$$\frac{\partial}{\partial r}\left(\frac{\partial^2 u_z}{\partial r^2} + \frac{1}{r}\cdot\frac{\partial u_z}{\partial r}\right) = 0|_{r=R_0} \quad .(10)$$

$$u_z = 0|_{r=r_0}$$

$$\frac{\partial u_z}{\partial r} = 0|_{r=r_0}$$

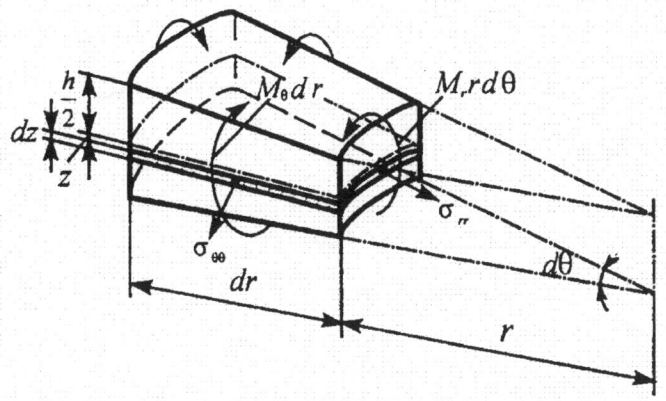

Fig. 49. To the calculation of the plate axissymmetrical stressed state

Equation system (10) gives enough relations for calculation (including the conditions of strain finiteness in the centre of the circular plate).

Solution (9) in view of (10) is located as stationary waves in the direction of r

$$u(r) = B_1 \cdot J_0(\lambda(f) \cdot r) +$$
$$+ B_2 \cdot Y_0(\lambda(f) \cdot r) + B_3 \cdot I_0(\lambda(f) \cdot r) + B_4 \cdot K_0(\lambda(f) \cdot r) \quad (11)$$

where $B_{1,2,3,4}$ are the coefficients dependent on edge conditions; $J_0(\lambda(f) \cdot r)$, $Y_0(\lambda(f) \cdot r)$, $I_0(\lambda(f) \cdot r)$, $K_0(\lambda(f) \cdot r)$ is the Bessel function;

$$\lambda(f) = R_0 \sqrt[4]{\frac{\rho \cdot h \cdot \omega^2}{D}}.$$

The graphs in Fig. 50 show the relative amplitude of the sensitive element displacement off the radius that depends on frequencies: 10 KHz (δ^{10}), 30 KHz (δ^{30}) and quasistatics (δ^0, the wavelength is considerably larger than the plate diameter). The graph clearly demonstrates the distinction of quasistatic calculations. The calculations were made for the following geometric sizes of the sensitive element: $R = 10$ mm; $r_0 = 2$ mm; $h = 0.8$ mm.

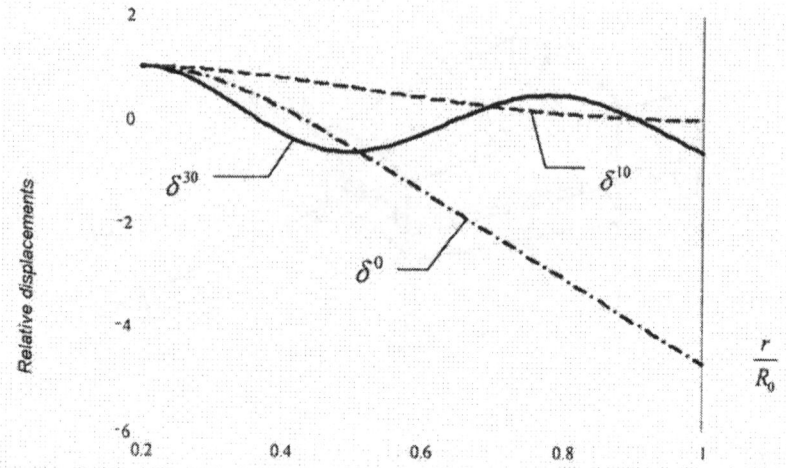

Fig. 50. Dependency of the sensitive element's displacements

9. Autoemission luminescent and explosive solid-state cathodes.

Testing and healing defects in MDM structures

Luminescence, one of the most vivid types of electromagnetic energy radiation, is used not only in computer or radiophone displays and lights. It occurs when almost all non-metallic media and minerals are exposed to powerful impulse electron beams.

Luminescence is used for the non-destructive testing of materials and includes: 1) recombination radiation, 2) radiation at optical charge transitions between the crystal zone and local impurity levels in the prohibited zone, 3) radiation at transitions in the "permitted" energy zones.

The manufacture of optoelectronic microstructures (hundredths of micron in size) uses electronic and ion lithography, as well as hard UV and X-ray radiation. Local testing of the composition and structure is based on these physical phenomena. The size of the area where an informative signal is registered is the same order of magnitude as the sizes of functional microcircuit elements. The known measurement techniques, such as scanning electron microscopy (SEM), X-ray spectrum microanalysis, secondary ion mass spectroscopy, sometimes cannot give precise local information. Optoelectronic materials can most informatively be studied by cathode luminescence (CL), excited by an electron [Stepovich M.A. Expanding the possibility of cathode luminescent microscopy for studying surface properties of direct bandgap semiconductors. J. Surface. X-ray, Synchrotron and Neutron Studies. 2002. No.7, p.97-103; Stepovich M.A. Assessing the precision of calculations of the distribution of minor charge carriers generated in a semiconductor material with an electron beam. Bull. of the Russ. Acad. Of Sci. Phys. 2003. Vol. 67. No.4. P. 602-606].

Thomas Alva Edison was the first to study electron emission in red-hot metal and carbon electrodes (1883) (thermoelectron emission). J.Fleming, a professor at London University, (who worked with Edison Electric Light Co.) in 1904 suggested using the Edison's effect to transform an alternating current into a constant current. He built the first electrovacuum diode, which contained a heatable negative cathode and a collector, the anode. In 1906 an American inventor Lee de Forest proposed a third, net electrode, for feedback. First triode-based radiocommunication devices were fabricated in the USA in 1912.

In 1901 a student of the British physicist Joseph John Thomson, the twenty-two-year-old Owen Richardson, a future Nobel Prize winner (1928), formulated the equation of the emission current

$$I = \overline{A}T^2 \exp\left[-\frac{e\varphi}{kT}\right],$$

where I is the current density in the saturation mode [A/cm2], φ is the electron work function [eV], T is the temperature of the emitter [K], k is the Boltzmann's constant (eV/K), $\overline{A} = 120.4$ [AK2/cm2] is the constant.

In 1903 Arthur Wenelt, professor at Berlin University (1871-1944), studied thermoelectron emission from precious metals and lo! discovered an increase by several orders. This anomaly was characteristic only of the wires that had lain on the marble windowsill of the laboratory. The wires were contaminated by marble (calcium carbonate), which in vacuum degraded to carbon oxide. A.Wenelt purposefully covered the metallic wire with an oxide layer to increase emission from the cathode. The thickness of the layer can be chosen by formula (1) in Appendix 4. The thickness of the oxide must be chosen deliberately.

There is a production technology of the MDM cathode, which may increase the current density by 5-7 times by covering its surface with a thin film of polypropylene (hydrocarbon molecules take a direct part in the formation of the through conductivity current) [Gyngazov S.A. Studying electrophysical properties of thin-film MDM systems under extreme environment conditions and electric fields. Ph. dissertation thesis. Tomsk: Tomsk State Academy of Control and Radioelectronic systems. 1995].

Contemporary optoelectronic devices are based on multi-layer oxide cathodes made of large-grain metal powders (nickel) with fine and coarse metallization. The technologies and constructions of cathodes, as well as the analysis of physico-chemical processes in electrovacuum devices can be found in [67].

Autoemission cathodes (AEC) are usually based on tungsten and transition metals, such as chrome, niobium, silicon and germanium. In vacuum they degrade, i.e. the curvature radiuses of the emission tips on the working surface increase. It reduces the local electric field strength and essentially decreases the autoelectron current. The tips are blunted due to bombardment by ions in residual gases. Ion formation has an avalanche nature (ion shock ionization).

Carbon autoemission cathodes (CAEC), made of carbon microfibers and oriented pyrolithic carbon (reactor graphite), have high mechanic strength and stability towards ion bombardment. The fibers are 7-10 µm in diameter and can be used in multi-tip "cold" cathodes.

Among the known materials, sulphide luminophors are the most promising. This result was obtained experimentally. Calculations showed that the maximal permissible heat load q_0 on a luminophor must not exceed 0.1 W/cm2. The temperature of the luminophor will be not higher than 100°C.

In view of the limit on the current load of a single carbon fiber (20÷30 µA), it was demonstrated that in the diode configuration of an elementary light source the area of the light spot should not exceed 0.33 cm2. The performance is approximately 20%.

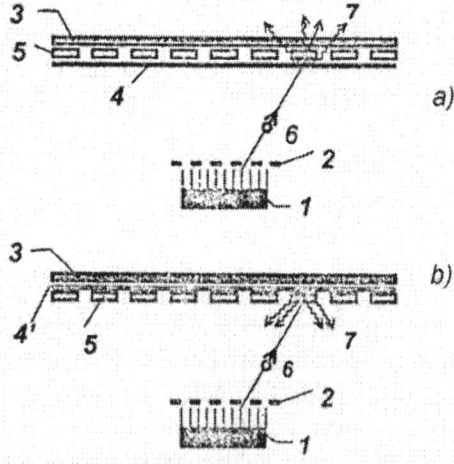

Fig. 51. Two layouts of cathodeluminescent light sources (1 is the cathode; 2 is the modulator; 3 is a fragment of the glass plate of the light source; 4 is the non-transparent mirror coating; 4 'is the transparent conducting coating; 5 is the luminophor, 6 are the electrons, 7 is the light).

Cold emitters based on silicon oxides, wide-band semiconductors (diamond, AlN, AlNiMo) under certain conditions have autoemission properties. The most stable emission occurs in tip cold diamond cathodes [Belyanin A.F. et al. Diamond and Related Materials. 1999. Vol. 8, p. 369-372].

Testing and defect healing in MDM structures

We have formulated a physical model of the autowave processes that take place at the initial reversible stage of the electric breakdown of the DE thin layer (this is realised at the photoelectric effect in submicron field transistors). Random fluctuations of the electromagnetic field, caused by structural defects, increase in DE due to positive feedback (by injection current and local temperature rise). Electric instability of quasineutral solid-state plasma (near micropeaks on the electrode surface) causes current oscillations and sometimes electromagnetic radiation. The frequency of electromagnetic wave radiation equals the frequency of absorption. This information can be used for the non-destructive testing and "healing" of technological defects.

A successful combination of electrophysical, chemical and mechanical properties of Si-SiO2 systems makes them a good choice for micromachining. These systems are the basis for electret sensors and actuators (microphones, pressure sensors, micropumps). The main requirement for electrets is the stability of their properties, which depends on the presence of defects in the crystal lattice (first and foremost, near the interfaces) [Kozodayev D.A. Electret effect in Si-SiO2 and Si-SiO2-SiN4 structures. Ph. dissertation thesis. St.Petersburg: Lenigrad Electrotechnical Institute, 2002]. This section determined working frequencies of the plasmochemical technique for healing defects in MDM structures. These results may be used for controlling parameters of various solid-state electronic devices based on silicon and its oxides, as well as gallium arsenide.

The properties of metallized DE depend on the technology of the preparation of their surface, the technique of electrode placement and the structure of materials. Scientific research experimentally established the essential part played by Joule heating of materials near the metal-DE interface and injection current through DE [Andreyev V.V. Physics and chemistry of material processing. 2001. No.6, p.47; Kropman D.I. et al. Russian Microelectronics (Mikroelektronika). 1986. Vol.15. No.4, p.376]. In view of these factors, this book formulates a hydrodynamic model of the electron-phonon system and determines the electromagnetic field frequency which makes "healing" structural defects possible.

Decreased sizes of diodes and transistors (IC elements), as well as the surface roughness of electrodes, lead to a local increase in the electric field strength in a relatively thin DE layer, bringing it closer to injection. E.g., in field transistors the injection of free charge carriers into the sub-gate DE becomes their operating mode.

Due to electron emission from the cathode, thin DE layers in contact with metal exhibit semiconductor properties. This gives one an opportunity to control the electrophysical characteristics of MDM structures by an external field due to the changing electroconductivity of DE.

At an internal electric field strength $E_z \geq 2.5$ kV/cm and rather high local temperature (300-600°C) near micropeaks on the electrode surface the movement of free charge carriers in the field of DE crystal lattice defects is determined by their drift, diffusion and recombination. A combination of these dynamic unstable processes brings about autooscillations in the electron-phonon system. Experimental data explain the leading role of electron injection into DE from metal electrodes (the recombination of electrons and holes being secondary).

Measuring electroconductivity is one of the simplest and most precise methods for studying defects in DE and high ohmical semiconductors [Vladimirov V.I., Lupashku R.G. Strength of Materials. 1973. No. 4, p.70; Ridley B.K. J. Appl. Phys. 1975. Vol.46. No.3, p.998). Microdefects in DE are identified by measuring the oscillation frequency of the current and electromagnetic radiation which occur at the initial reversible stage of the electric breakdown in MDM structures under a constant external electric field (or laser radiation) [Bogomol'nyi V.M. Measurement Techniques. 2002. No.10, p.46; Golovin Y.I. et al. Condensed Media and Interfaces. 2002. Vol.4. No. 1, p.5; Sakurada T., Kadoya Y., Yamanishi M. Jpn. J. Appl. Phys. 2002. Vol. 41. Pt. 2. N 3A, p. L-256].

The dynamics of formation and development of mesoscopic structural defects can be studied by the electromagnetic emission technique, which registers their own radiation, occurring at elastic strain and degradation in crystals). In the areas where electrons are injected into DE, there are channels of high conductivity, or "current filaments", which at local temperature increase are filled with solid-state plasma. The flow of electrons injected into DE, compensated by the space charge of the crystal lattice, forms a nonlinear dynamic electron-phonon system, in which random fluctuations of the electromagnetic field (connected with structural defects) increase. This causes current oscillations under an external constant electric field.

If there is a time-dependent disturbance of the electromagnetic field in a crystal, the uneven space distribution of charge carriers is in accord with its non-stationary disturbance. The simultaneous realization of the time and space resonances is one of the reasons why out of an infinite number of combined waves only one space stationary wave is excited effectively. The equilibrium space distribution of charge carriers does not lag behind the disturbance, which changes in time if the low-temperature solid-state plasma is quasineutral.

In "cold" plasma, when the characteristic velocity of the chaotic heat movement of free electrons is essentially smaller than the phase velocities of electromagnetic waves, one can use the hydrodynamic model based on averaged equations of the movement of charge carriers. For the disturbances of the free charge carrier concentration to travel as electrohydrodynamic waves, we need to meet the condition

$$\omega_c < \omega < \omega_d \qquad (1)$$

$$\omega_c = \frac{en\mu}{\varepsilon_0 \varepsilon_{33}^T}, \quad \omega_d = \frac{c_0^2}{D}, \quad D = \frac{\mu kT}{e}, \qquad (2)$$

where n is the concentration of electrons injected into DE, e is the electron charge, μ is the electron mobility, ε_0 is the absolute dielectric permeability of vacuum, ε_{33}^T is the relative dielectric permeability of the crystal, k is the Boltzmann's constant, T is the temperature, ω is the eigenfrequency of the electron-phonon system, ω_c is the frequency of the Maxwell's dielectric relaxation, ω_d is the diffusion frequency of charge carriers, c_0 is the velocity of the waves of the electromagnetic field disturbances, D is the diffusion coefficient. The disturbance should have no time to relax within the oscillation period, but it may travel farther than its wavelength. The uneven distribution of current carriers creates the electric field E, which satisfies the Poisson's equation

$$\nabla E = e(n - <n>)/\varepsilon_g, \qquad (3)$$

where ∇ is the Laplace operator, $\varepsilon_g = \varepsilon_{33}^T \varepsilon_0$ is the absolute dielectric permeability (F/m).

The electroconductivity equation in view of electron diffusion (2) has the form

$$J = \sigma E - eD\nabla n = \sigma E - e\mu A \nabla n, \qquad (4)$$

where $A = kT/e$, $\sigma = e\mu < n >$ is the conductivity. The current function J satisfies the continuity equation

$$e\dot{n} + \nabla J = 0. \quad (5)$$

Substituting (3), (4) into (5), we obtain an equation for the function $n(x,t)$, which determines the condition of plasma quasineutrality [Kornyushin Y.V. Transfer Phenomenon in Real Crystals under External Fields (diffusion kinetics of current carriers and microdefects). Kiev: Naukova Dumka, 1981].

$$\dot{n} + e\mu < n > \frac{n - <n>}{\varepsilon_g} = A\mu\nabla n. \quad (6)$$

At the initial moment of time we take into account electron distribution

$$n(x,0) = <n>(1 + m\cos\kappa x). \quad (7)$$

In view of (7) the solution of equation (6) has the form

$$n(x,t) = \{n\}[1 + m(\cos\kappa x)\exp(-t/\tau)]. \quad (8)$$

The setting time of the equilibrium distribution of charge carriers τ (response to outer influence) must be smaller than the characteristic time of the change in disturbance Δt ($\tau < \Delta t$). At the moment of time $t = 0$ charge carriers are distributed according to the law $n(x,0)$. If at $t > 0$ disturbance ceases, the space distribution of charge carrier concentration gradually levels out.

Substituting (8) into (6), we determine the relaxation frequency of electron energy $f_e = 1/\tau$

$$\frac{1}{\tau} = \frac{e\mu < n >}{\varepsilon_g} + A\mu\kappa^2. \quad (9)$$

From (9) the condition of the equilibrium in the electron-phonon system at $\tau < \Delta t$ follows from the equation

$$\frac{4\pi^2 A\mu}{v_{ac}^2} f^2 - f + \frac{e\mu < n >}{\varepsilon_g} > 0, \quad (10)$$

where $f = 1/\Delta t$ is the disturbance frequency, $v_{ac} = 2\pi f/\kappa$ is the speed of acoustic wave propagation.

Taking into account the expression $v_{ac} = \dfrac{2\pi f}{\kappa}$, the second component in formula (9) gives the relaxation frequency of electron energy

$$f_{min} = \frac{e v_{ac}^2}{4\pi^2 \mu kT}.$$

At local increase in temperature and electric field strength near roughnesses on the electrode surface, one must also consider the effect of heating free charge carriers by an electric field, when the electric field strength satisfies the inequality

$e < E_z > l_e \geq kT$.

At 300K and $l_e = 10^{-6}$ cm "hot" electrons appear in fields E_0

$E_0 = \dfrac{kT}{el_e} = 2500$ V/cm.

This estimation corresponds to experimental data. The paper [Shur M. Electron. Letters. 1976. Vol. 12. No.23, p.615] contains an experimental study of operating modes of n-Ga-As-based diodes at electric field strengths $(1.4 \div 2.5) \cdot 10^4$ V/cm. For n-GaAs crystal at $T = 300$ K with $\mu = 4000$ [cm^2/(V$\cdot$s)], $v_{ac} = 5.2 \cdot 10^5$ [cm/s], taking into consideration that $kT/e = 0.025$ V and we estimate $f_{min}^{(1)} = 68$ MHz.

Writing equation (10) in the general form and putting to zero its right-hand part, let us determine the excitation frequency of electrohydrodynamic waves in the n-GaAs diode, discounting electron injection from the electrode. We shall use the solution of the quadratic equation

$af^2 - f + b = 0$,   (11)

where $a = \dfrac{4\pi^2 A\mu}{v_{ac}^2}$, $b = \dfrac{e\mu <n>}{\varepsilon_g}$.

In some practical applications $ab \gg 1$, and the frequency f can be calculated by the formula $f = (b/a)^{1/2}$.

At $T = 300$ K, $<n> = 10^{23}$ m^{-3}, the above-mentioned μ, v_{ac} and $\varepsilon_g = 11 \cdot 8.85 \cdot 10^{-12}$ F/m, from the solution of equation (11) we receive $f_{min} = 8.24$ GHz, which corresponds to the experiment [Janes D.B. et al. J. Appl. Phys. 1995. Vol.78. No.11, p.6616], where the frequencies of a n-GaAs-based diode were measured in the range of 45 MHz – 20 GHz.

Charge carrier injection, controlled by an external field, is a relatively "mild" influence, more preferable compared to ion implantation, which causes hard-to-anneal defects.

By analysing the spectra of electron paramagnetic resonance (EPR), researchers [Kropman D.I. et al. Russian Microelectronics (Mikroelectronika) 1986. Vol.15. No.4, p.376] studied the electrophysical and chemical processes of thermal treatment of a Si-SiO2 system. It was found that the Si-SiO2 interface plays the getter role for hydrogen. The concentration of electrically active defects decreases, and the imperfections of material structure are eliminated.

Current oscillations (in a circuit with an MDM sample) and electromagnetic radiation from DE can be rather precisely measured by standard radiotechnical methods. This information can be used to identify the slightest structural microdefects at their own resonance frequencies, which can be calculated by the given formulas.

Molding of thin-film MDM structures

Under a constant external electric field at the initial stage of the electrothermal breakdown of SiO2 there are channels of electron conductivity, which are the electron emitters in vacuum electronic devices ("cold cathodes"). Molded MDM structures can be used in flat matrix addressing displays. Their technology is simple and compatible with IC technology. "Cold cathodes " are inexpensive and work at up to 2 V [Kramer S.S., Haskelberg M.B. Studying the molding and degradation of emission parameters of MDM systems. Soviet Physics Journal. 2000. No.7, p.7-13].

Similarly to the degradation of thin-film MOS systems, at the initial stage of the electric breakdown MDM structures generate neutral electron traps in the DE volume. The generation is caused by the release of atomic and molecular hydrogen and atomic oxygen from the DE zone near the anode and its movement through DE to the cathode-DE interface. When the concentration of defects in a locality near the cathode-DE interface becomes critical, we observe an electrothermal breakdown and the formation of molded channels (MC). It has been found that the upper electrode breaks away above an electrically active structural macrodefect in DE.

The experiments analysed Al-SiO-Ni MDM systems. The SiO layer was 35 nm thick, the metal electrodes, 20 nm and 0.5-1.0 μm accordingly. At 0.5-2.0 V the current begins to grow irreversibly in time (the first molding stage). The temporal dependence of current follows the law $I \cong t^{1/2}$. This behaviour is characteristic of the transport of diffusion-limited charges (atomic and molecular hydrogen and atomic oxygen).

Displays with strictly periodical MC contain electrodes with a regular system of micropeaks on the electrode surface in the shape of truncated pyramids 0.1-0.2 μm high. These microtips are created when the DE film is about 40 nm thick. At ~80 nm it is not possible to mold an MDM structure with a smooth electrode. The electric field strength due to the strengthening of the thin film is about ~104 V/cm [Vorobyov G.A. et al. Physico-technological aspects of fabricating MDM cathode matrices for a flat vacuum luminescent screen. J.Communications Technology and Electronics. 1992. Vol. 37, No. 4, p.413-420].

Experiments and calculations have shown the effect of the microrelief roughness of the electrode surface on the electric molding of an MDM structure. This causes a considerable (by dozens of times) change in local energy emission as compared to the smooth electrode.

In thin-film metal-silicon dioxide and silicon oxinitride-aluminum systems molding occurs under normal atmospheric conditions at $E_{form} \cong 1$ kV/mm. The diameter of the channel on the DE surface is ~15-25 nm. It has been calculated that the temperature at the outer edge of MC is 2-3 times lower than the maximal temperature in its central area, by 150-250°C higher than the electrode temperature, and drops to the electrode temperature when put back at a distance comparable to its diameter.

The MC diameter $d = 15$ nm indirectly characterized the diameter of the spherical surface of the micropeak top on the electrode surface [Vorobyov G.A., Gaponenko V.M Peculiarities of electric molding and breakdown of a thin-film MDM system. Soviet Physics Journal. 1992. No.1, p.58-63].

MDM structures (silicon oxides and nitrides, anode niobium, aluminum and tantalum oxides) are used in memory elements, condensers and piezo-electronic devices. Near weak fields the DE breakdown (in tantalum and silicon oxides) is determined by "hot" electrons. The current in MDM structures grows in time [Laleko V.A., Ershova N.Y., Dragan I.I. On the mechanism of breakdown in anode tantalum oxide. Sovjet Solid State Physics. 1992. Vol.34. No.7, p.2118-2121]. Near strong electric fields the breakdown happens due to space charge accumulated in the DE. The field maximum is near the non-injecting contact (for the anode). The breakdown begins when the current activation energy reaches $\Delta E_{br} \cong 0.1$ eV.

For MNOS structures under a constant external electric field we observe the strengthening effect, characterized by an increase (by several times) in the time of the electric breakdown.

The industrial development of cathodoluminescent displays (flat vacuum fluorescent and field-emission displays) determined the necessity to study the effect of electroconductivity, the size of the crystallophosphorus grains and screen thickness on the image brightness. It has been found that the optimal screen thickness is 5-10 μm, with a grain size of 2-3 μm (2-3 layers) [Strel'tsov A.V. Physico-chemical processes in luminophor layers of flat cathode luminescent displays. Ph. dissertation thesis.. Saratov: Saratov State University. 2002; Strel'tsov A.V., Bukesov S.A. Temperature field simulation in low-voltage vacuum displays. Solid State Electronics. 2001. Vol. 45. No. 6, p. 887-892].

Explosion-emission cathodes are used in smart technologies and accelerating equipment. Diode-type electron sources form microsecond electron beams at 100-600 kV, and low-energy beams with an energy of 10-25 keV. Fields in such diodes are strengthened by carbon-fiber sources of cathode plasma and DE inclusions [Korenev S.A. Diod-type sources with high-temperature conductor cathodes and anodes. Dubna: The United Institute for Nuclear Studies 1994. P13-94-192; Mesyats G.A. Impulse energetics and electronics. Moscow: Nauka. 2004]. In 2004 Russian scientists were the first to use this principle to create a laser impulse which equals an atomic plant in power.

The reduction effect of the emission threshold at covering the cathode with a DE film was established for tip Si-based cathodes (filament crystals with ~100 μm height and curvature radius at the top <25 nm, molded by the "vapour-liquid-crystal" technique) with 0.1-2 μm AlN (Fig. 52) [Belyanin A.F. et al. Diamond and Related Materials. 1999. Vol. 8, p.369-372]. The autoemission of single tip cathodes was measured in a vacuum camera under 1.33·10-6 Pa. The distance between the cathode and the anode was ~200 μm.

 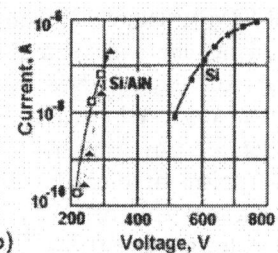

Fig. 52. (a) Electron-microscopic image of a tip Si/AlN cathode. (b) CVC of tip non-heatable Si and Si/AlN cathodes.

Tip cathodes are produced by lithography, which limits the possibility of creating emission centres (tips) of high density in a wide area. In some experiments the high density (up to 2.5·109 cm-2) of emission centres in the area >200×200 mm was achieved by 3-dimensional opal matrices (up to 10 layers), molded on plates of Si, pyroceramics and melted quartz (Fig. 53). Opals are densely packed (mainly by the cubic law) nanospheres of silicon oxide (very close in diameter), the size of which in different samples may vary from 200 to 600 nm. The regular packaging of silicon oxide nanospheres forms a 3-dimensional lattice with a given periodicity, which may be characterized as a 3D superlattice, and the system on the whole as an "optical " or "photon " crystal. At the given diameters of spheres these packages contain structural cavities of 160-400 nm, which can be partially or completely filled with semiconductor, superconducting, optically active, magnetic and other materials. Thus, in the opal matrix there will be a 3-dimensional superlattice of filling particles (the size of clusters may vary from 10 to 180 nm). It is by this method that the first 3-dimensional nano-composites with a characteristic nanoscale structure were obtained [Samoilovich M.I. et al. Nanotechnology. 2002. Vol.13, p.763-767].

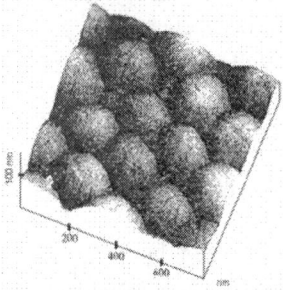

Fig. 53. Surface of an opal matrix on a pyroceramic plate, shown in an atom-force microscope image

The micropeaks of SiO2 spheres (on the surface) were sharpened by magnetronic spraying layers of an emitting material (diamond-like carbon or DLC, AlN, layer-like structures) by the HF technique. It is also possible to use intermediate passive metal or DE layers with an emitting layer. The sharpening effect is connected with a change in the relative speeds of the formation of film parts, situated under different angles to the spraying source. Depending on the thickness of the deposited layer, the shape of the surface obtained by sharpening the SiO2 sphere may vary (Fig. 54).

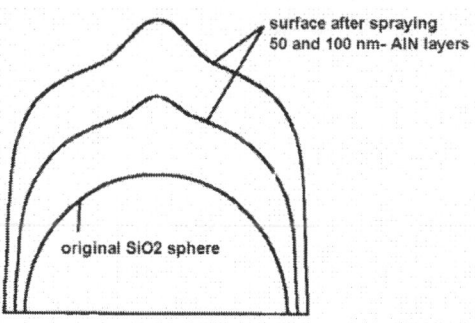

Fig. 54. Diagram of changing the surface configuration of the SiO2 sphere when depositing a AlN film

Varying the value and sign of the electric voltage on the plate holder U , experimenters changed the degree of crystallinity and the crystal phase structure of various films. For example, the structure of AlN films changed from disordered to a texture with a crystallite disorientation angle <0.5°. At U = −50 V amorphous AlN (AlNam were steadily obtained ; at U = +45 V polycrystal AlN (AlNcr) films were formed, with crystal phase content >50 vol.%. To reduce the emission threshold energy of planar cathodes, the growth surface of diamond-like carbon (DLC) films by HF magnetronic reactive spraying was covered by AlN films (crystal and amorphous, AlNcr и AlNam); AlN films, alloyed by Mo (AlN:Mo); MgOx; Al2O3; ZnO and GaAs with a ~10 nm thickness. The plates with a diamond-like carbon layer were immobile in the process of DE film deposition.

There are solutions where carbon films are obtained by spraying a graphite target with an ion beam. The concentration of the diamond phase was 1-10 vol.%. We have also used X-ray amorphous films of diamond-like carbon, molded by the glow discharge technique. The deposited AlNcr, AlN:Mo, MgOx and Al2O3 films contained up to 40 vol.% of the crystal phase, the crystallites of which were axially texturized by <0001> (for AlN and Al2O3 films) and by <111> (for MgOx). Apart from these materials, AlNam and GaSb X-ray amorphous films were also used.

The emission characteristics of planar cathodes, based on 3D opal matrices, carbon films or layer-like DLC/DE structures were measured in the impulse mode on a specially designed vacuum device under $1.33 \cdot 10{\text -}3$ Pa. The gap between the cathode and the anode in the process was $\Delta = 160 - 500$ μm (Fig. 55). The electric field strength in the gap between the anode and the cathode (substrate) is $E = U / \Delta$, where U is the potential difference between the electrodes.

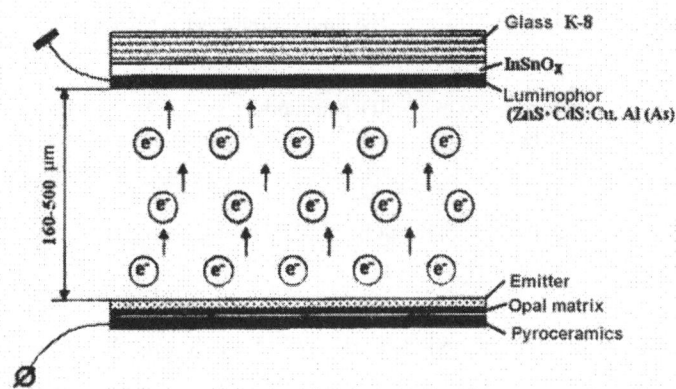

Fig. 55. Layout of the device for measuring CVC and estimating the number of emission centres of non-heatable cathodes based on the layer-like structure "substrate /SiO2/emitting layer

We consider the emission characteristics of planar non-heatable cathodes with a layer of ordered SiO2 spheres 200 nm in diameter (opal matrix). Emitting materials were DLC, AlN, and layer-like structures DLC/AlN; DLC/ZnO; DLC/MgOx; DLC/Al2O3; DLC/AlN:Mo, etc.

Opal composites can help to control the structural characteristics of photon bands to obtain 3-dimensional nanocomposites with unusual, e.g. quantumoptical properties. Such systems are of practical interest because they can give spontaneous optical emission or localize and control light. Thus, these optically active media can be used in laser gyroscopes or in laser reading devices for CDs, used in computer, audio and video equipment.

Table 8 contains information about the composition and possible applications of 3D-nanocomposites based on opal matrices [Romanov S.G. et al. On the possibility of controlling optical properties of 3D "soft" photon opal-based crystals. Sovjet Solid State Physics. 1996. Vol. 38. No.11, p.2051-55].

Type of filling	Composition of filling	Physical model of the nanocomposite	Possible application
Semicon-ductor	CdS, GaAs	Seebeck's 3D nanostructures	Thermoion energy converters
	GaAs, CdTe, HgSe, Te	3D superlattices of Shottky diodes and similar elements	Semiconductor nanoelectronic devices
Supercon-ductor	In, Pb Possibly high temperature su-	3D superlattices of Josephson contacts (transitions)	Generators and amplifiers of gigaherz electromagnetic

	perconductivity		waves (>10-20 GHz)
Optically active me-dia	ZnS, ZnO	3D superlattices of photodiodes	Elementary particle counters
	High-refractive materials	3D nanooptical sys-tems	Active control ele-ments in laser sys-tems

Table 8. Compositions and applications of 3D-nanocomposites based on opal matrices

The formula in § 3.8 can give the estimation of the electric field strength concentration near the tip Si cathode. At $H = 100$ µm, $R = 2.5 \cdot 10^{-2}$ µm $E_z^{max} = \sqrt{H/R}\rangle E_z\langle = 63\rangle E_z\langle$, where E_z is the integral electric field strength (averaged by the width of the gap between the electrodes).

10. Physical models of porous DE and semiconductors

Porous polar DE and semiconductors have a very developed surface and are therefore used in chemical sensors for testing gaseous and liquid media. Experiments have shown that porous piezoceramics has better electromechanical properties than monolithic ceramics. At certain geometrical ratios and pore shape the piezomodule d_{33} increases and d_{31} is almost zero, which increases the PZ sensitivity of sensors.

Porous silicon (PS) is used for local insulation in IPOS (Isolated by Porous Oxidized Silicon) integrated circuits. This led to the appearance of industrial silicon-on-insulator technologies (SOI).

PS is obtained from monocrystal silicon by anode electrochemical treatment of its surface in water solutions of hydrofluoric acid (HF).

The discovery of visible photoluminescence of PS at room temperature (1990) excited interest in its optoelectric application. PS is the basis for various solid-state electronic devices (in measuring and computer technology, in medicine).

Lately there has been a lot of research into nanosized effects in PS, e.g. visible luminescence (photo- and electroluminescence) caused by photoexcitation and injection of non-equilibrium charge carriers from solid-state and electrolytic contacts) [Properties of porous silicon. Ed. I. Canham.: EMIS Data reviews. – 1997. No.18; Zimin S.P., Komarov E.P. Analysis of dielectric permeability of porous silicon in the frame of a two-phase model. Soviet Electronics Journal. 1998. No.3, p.48-51; Zimin S.P. Concentration of charge carriers in the porous silicon monocrystal matrix. Letters to JETP. 1995. Vol. 21. No.24, p.46-50; Zimin S.P. et al. Charge carrier transfer processes in thin-film PS structures. Soviet Electronics Journal. 2000. No. 1, p.15-20].

The analysis of the mechanical stress-strain state near holes in preloaded solid materials has shown that in the case of closely situated holes there is a decrease and a considerable increase (by 2-4 times) of the concentration of stresses in comparison with isolated holes, and the correction in view of physical and geometrical nonlinearity may reach 40-90% [Levin V.A., Lokhin V.V., Zingerman K.M. On a technique for calculating effective characteristics of porous bodies at finite strains. Mechanics of Solids. 1997. No.4, p.45-50].

Electroelasticity relations for piezoelectric composite materials are given in [Sokolkin Y.V., Pan'kov A.A. Electroelasticity of piezocomposites with irregular structures. Moscow: Nauka. Fizmatlit. 2003; Pan'kov A.A., Sokolkin Y.V. Solution of the boundary value problem of electroelasticity for piezoactive composites by the periodic components technique. Mechanics of Composite Materials and Constructions. 2002. Vol.8. No.3, p.365-384; Pan'kov A.A., Sokolkin Y.V. Effect of geometry of ellipsoid pores on the properties and distribution of deformation fields in piezoceramics. Mechanics of Composite Materials and Constructions. 2003. Vol.9. No.10, p.87-95].

Electrochemical technology of PS production can use mathematical models of randomly chaotically located spherical (CLS) inclusions and pores [Brathia S.K., Perelmutter D.D. A random pore model for fluid-solid reaction. I: Isothernal kinetic control. AIChE Journal. 1980. Vol.7, p.379-385]. The formulated model of CLS cavities takes into account the time variation of the characteristics of porous material. The system of equations includes the kinetics equation of the material's dissolution.

Calculation of specific surface and chemical reactivity of porous materials

Specific surface is important for describing the kinetics of the leaching of minerals dispersed in the inert ore matrix. Most often minerals dissolve in the whole volume of ore particles, so the specific surface is a time- and space-variable value. The known recommendations to calculate this characteristic for hydrometallurgical processes seem too simplified. Let us use the experience of modeling the structure of porous materials, porous catalysts in particular.

Solid porous bodies consist of a frame, sometimes called a skeleton, and a system of cavities, the pore space. It is common to divide porous materials into corpuscular and spongeous. In corpuscular materials, pores are formed by cavities between compact solid particles, which make up the skeleton. In spongeous materials, pores are channels and cavities in the solid.

For the components of the space with a unit volume the obvious identity is true

$$\kappa + \varepsilon = 1, \quad (1)$$

where κ is the part of volume occupied by the frame; ε is the cavity part.

It is more convenient and precise to model corpuscular porous materials by a system of solid particles with a definite geometry (plates, cylinders, spheres). Spongeous media can be reproduced more easily by a system of pores, or cavities, with the oval form given in the common case.

Let us state the basics of the stochastic corpuscular model of porous materials. The mineral frame of the material is formed by a system of corpuscles, chaotically distributed in space. They are given a definite geometrical shape, e.g. that of a sphere. In the unit volume we take the average density n and the function of the density of distribution of corpuscles by size.

The radius of corpuscle spheres is taken as equal to the constant value of ρ. At high density the spheres may overlap. The radius ρ and density n are the main parameters that determine the structural characteristics of CLS and inclusions.

The porosity and specific surface depend on the dimensionless parameter

$$\alpha = \frac{4}{3}\pi\rho^3 n . \quad (2)$$

Let V be the allocated volume of the porous medium, much larger than the volume V_0, which does not contain any corpuscle centre. Obviously, the probability of the randomly chosen corpuscle sphere being outside the volume V_0 is $(V - V_0)/V$. Hence, from the condition of the non-correlation of the distribution of the sphere centres, the probability of nV corpuscles being in the volume $(V - V_0)$ is

$$p(V_0) = (1 - V_0 / V)^{nV} = (1 - nV_0 / nV)^{nV} .$$

Since the number nV is enormous, and the product nV_0 is limited, at the limit the probability that the volume V_0 contains no sphere centres is

$$p(V_0) = \exp(-nV_0) .$$

Hence, taking into account formula (2) at $V_0 = \frac{4}{3}\pi\rho^3$, we have

$$\varepsilon = \exp(-\alpha) . \quad (3)$$

The surface area of non-overlapping CLS-corpuscles in the unit volume of the porous medium is $4\pi\rho^2 n$. When the spheres overlap, the area is smaller. In view of the fact that a randomly chosen point on the surface of the sphere is not covered by another sphere with the probability ε, we obtain the final expression of the specific surface

$$S = 4\pi\rho^2 n\varepsilon = [3\alpha\exp(-\alpha)]/\rho. \qquad (4)$$

Formulas (1)-(4) constitute the corpuscular model of a porous body.

The model of CLS-cavities is also determined by the cavity radius ρ and cavity number n in the volume unit. According to expressions (1), (3), porosity now is

$$\varepsilon = 1 - \exp(-\alpha). \qquad (5)$$

Let the function of the density of the spherical cavity radius distribution meet the condition of normalizing

$$\int_0^\infty f(\rho)d\rho = 1.$$

We introduce the parameters

$$\rho_E = \int_0^\infty \rho f(\rho)d\rho; \quad S_E = 4\pi\int_0^\infty \rho^2 f(\rho)d\rho; \quad \varepsilon_E = \frac{4}{3}\pi\int_0^\infty \rho^3 f(\rho)d\rho,$$

to denote the integral radiuses, surface and volume of the system of spheres-cavities (per volume unit of the porous material) discounting the possible overlap of cavities. At high density n in such cavities overlapping volumes and surfaces are taken into account twice, three times, etc. If we consider overlaps, the factual specific volumes, surfaces and linear sizes should be determined by the following expressions:

$$\varepsilon = 1 - \exp(-\varepsilon_E), \qquad (6)$$
$$S = S_E(1-\varepsilon), \qquad (7)$$
$$\rho = \rho_E(1-\varepsilon). \qquad (8)$$

We obtain the expressions for S_e, ε_E in the form of analytic dependences.

From the equation of the reaction kinetics

$$\frac{d\rho}{dt} = kC \qquad (9)$$

and the balance equation

$$\frac{\partial f}{\partial t} + \frac{\partial}{\partial \rho}\left[f\frac{d\rho}{dt}\right] = 0$$

we find

$$\frac{\partial f}{\partial t} = -kC\frac{\partial f}{\partial \rho},$$

where k is the constant of the reaction speed; C is the concentration of the reagent.

If all cavities of the medium have the same radius, $f(\rho)$ is represented by a delta function. The derivative in this case $d\rho_E / dt$ equals the constant value:

$$\frac{d\rho_E}{dt} = \int_0^{\infty}\left[f(\rho)\frac{\partial \rho}{\partial t}d\rho + \rho\frac{\partial f}{\partial t}d\rho \right] =$$

$$= kC\int_0^{\infty} f(\rho)d\rho - kC\int_0^{\infty}\rho\frac{\partial f}{\partial \rho}d\rho = 2kC \qquad (10)$$

After multiplying expression (9) by $4\pi\rho f(\rho)d\rho$ and integrating we obtain

$$\frac{dS_E}{dt} = 4\pi kC\rho_E. \qquad (11)$$

After multiplying expression (9) by $(4/3)\pi\rho^2 f(\rho)d\rho$ and integrating we come to the equation

$$\frac{d\varepsilon_E}{dt} = \frac{kC}{3}S_E. \qquad (12)$$

After integrating equation (11) by time, taking into account formula (10), we have

$$S_E = S_{E_0} + 4\pi[\rho_{E_0} kCt + (kCt)^2].$$

Using formulas (6), (7), we find

$$\frac{S}{S_0} = \frac{1-\varepsilon}{1-\varepsilon_0}\left[1 + 4\pi kCt\left(\frac{\rho_0}{S_0} + \frac{1-\varepsilon_0}{S_0}kCt \right) \right]. \qquad (13)$$

Similarly, one can derive from equations (12) and (13)

$$\varepsilon_E = \varepsilon_{E_0} + \frac{1}{3}kCt\left[S_{E_0} + 2\pi\rho_{E_0}(kCt) + \frac{4}{3}\pi(kCt)^2 \right].$$

The latter equation in view of (6) is transformed into

$$\varepsilon = 1 - (1-\varepsilon_0)\exp\left\{ -\frac{1}{3}kCt\left[\frac{S_0}{1-V_0} + 2\pi\frac{\rho_0}{1-\varepsilon_0}(kCt) + \right. \right.$$

$$\left. \left. + \frac{4}{3}\pi(kCt)^2 \right] \right\} \qquad (14)$$

Formulas (13), (14) are the main when calculating variables in the time of specific reactive surface and porosity.

Fig. 56 features the dependence of the specific surface of the reaction on porosity, built from expressions (13), (14).

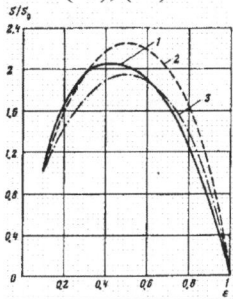

Fig. 56. Dependence of the relative reactive surface on the porosity of material. Models: 1 is based on [Brathia S.K., Perelmutter D.D. A random pore model for fluid-solid reaction. I: Isothernal kinetic control. AIChE Journal. 1980. Vol. 7, p.379-385]; 2 is for CLS corpuscles; 3 is for CLS cavities.

As follows from Fig.56, at a porosity within $0.4 < \varepsilon \leq 0.6$ one can observe the maximal reactivity. This theory can be used to design chemical PZ sensors (with maximal sensitivity).

11. Calculation of piezotransducers at harmonic excitation

The temperature of a thin piezoceramic shell is determined with regard to dielectric, mechanical and piezoelectric losses; the comparison of the calculation results with experimental data show that error does not exceed 15%.

The electrophysical properties of ferroelectrics used in precision measurement devices, such as dielectric permeability, polarization, elastic and piezoelectric constants (especially near phase transitions), largely depend on temperature.

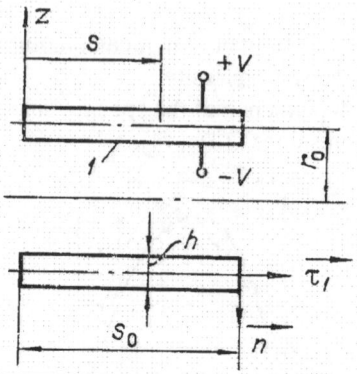

Fig.57. Analytical model of a transducer (1 is the thyristor)

We consider a cylindrical shell $0 \leq s \leq s_0$, with electric potentials $V = V_0 e^{i\omega t}$ on its outer cylindrical surface (Fig. 57).

The equations of the movement of the cylindrical shell have the form [Love A.E.H. A Treatise on the Mathematical Theory of Elasticity. N.Y.: Dover Publ. 1944]

$$\frac{dN_s}{ds} = -h\rho\omega^2 U_s, \quad \frac{dM_s}{ds} = Q_s, \quad (1)$$

$$\frac{dQ_s}{ds} - \frac{N_\theta}{r_0} + q_n = -h\rho\omega^2 U_z. \quad (2)$$

All the notations are given at the end of the section.

We take the relations of electroelasticity for a thickness-polarized piezoceramic shell in the following form [Boriseiko V.A., Grinchenko V.T., Ulitko L.F. Electroelasticity relations for piezoceramic rotation shells. Sov. Appl. Mechanics. 1976. Vol.12. No.2, p.26-33]:

$$N_s = D_N(\varepsilon_s + \nu\varepsilon_\theta - E_0), \quad N_\theta = D_N(\varepsilon_\theta + \nu\varepsilon_s - E_0),$$

$$M_s = D_M\kappa_s, \quad M_\theta = \nu M_s, \quad E_0 = (1+\nu)d_{31}E_z^{(0)}, \quad (3)$$

$$D_N = \frac{h}{s_{11}^E(1-\nu^2)}, \quad D_M = \frac{h^3\gamma}{12s_{11}^E(1-\nu^2)}, \quad E_z^{(0)} = -\frac{2V_0}{h},$$

$$\gamma = 1 + \frac{1+\nu}{2}\frac{K_p^2}{1-K_p^2}, \quad K_p^2 = \frac{2}{(1-\nu)}\frac{d_{31}^2}{s_{11}^E\varepsilon_{33}^T}.$$

The deformations of the middle surface of the shell are determined through displacements:

$$\varepsilon_s = \frac{dU_s}{ds}, \ \varepsilon_\theta = \frac{U_z}{r_0}, \ \kappa_s = -\frac{d^2U}{ds^2}, \ \kappa_\theta = 0. \quad (4)$$

Let us view the axisymmetrical oscillations by the first, lower mode. The edge conditions for equation system (1)-(3) will have the form

$$U_s \big|_{s=s_0/2} = 0, \ N_s = M_s = q_s = 0 \ \text{at} \ s = 0, \ s = s_0. \quad (5)$$

At certain values of the relations h/r_0 (as shown in [Love A.E.H. A Treatise on the Mathematical Theory of Elasticity. N.Y.: Dover Publ. 1944]) the function U_s^0 may be calculated by the "moment-less" theory of shells

$$U_s^0 = A \sin k\tilde{s} = \frac{U_s^0}{r_0}, \ \tilde{s} = \frac{s}{r_0}, (6)$$

$$A = \frac{(1-\lambda-v)E_0}{(1-\lambda-v^2)k \sin \dfrac{ks_0}{2}}, \quad k^2 = \frac{\lambda}{1 - \dfrac{v^2}{1-\lambda}}.$$

Using (6) from the main equation system (1)-(4), we find the other unknown functions, which determine the stress-strain state of the piezoceramic shell.

The equation of heat conductivity was calculated with the help of a computer. The calculations showed that the temperature is *constant along thickness*. The maximal growth of initial temperature is determined by the formula

$$T_{max} = \omega \sum_{i=1}^{3} \Phi_i / \alpha_t F .$$

The components of the function of the energy dissipation due to dielectric, mechanical and piezoelectric losses are calculated in the following way [Grinberg G.A., Kontorovich M.I., Lebedev N.N. On temporal behaviour of thermal breakdown. Sov. Physics – Technical Physics. 1940. Vol.10. No.3, p.199-216; Berlinkur D., Kerran D., Jaffe G. Piezoelectric and piezomagnetic materials and their application to transducers. In: Physical Acoustics. Ed. W.Mason. 1966. Part 1A, p.204-326; [15]. Unlike the solution of the unconnected problem of thermoelectroelasticity for an infinite cylindrical shell:

$$\Phi_1 = C_{el}V_0^2 \text{tg}_{de}, \quad (7)$$

$$\Phi_2 = \frac{0.7v(1+\text{tg}\delta_{mech})}{2(1+v)} \left[\varepsilon\varepsilon_s^2 + \varepsilon_\theta^2 + \varepsilon_z^2 + \frac{v(\varepsilon_s + \varepsilon_\theta + \varepsilon_z)^2}{(1-2v)} \right], \quad (8)$$

$$\Phi_3 = -E_z^{(0)} v[e_{33}''\varepsilon_z + e_{31}''(\varepsilon_s + \varepsilon_\theta)], \qquad (9)$$

where $\varepsilon_z = s_{13}^E \left(\dfrac{N_s}{h} + \dfrac{N_\theta}{h} \right) + d_{33} E_z^{(0)}$.

In the second formula, when one calculates the functions of heat dissipation, due to elastic and viscous elastic losses in piezoceramics, the compression strains ε_i ($i = s$, θ, z) and ε_i^2 are taken as positive, and the corresponding tension strains and their squares are considered negative.

When the geometry of a piezoceramic PZT-4 transducer, free from external mechanical forces, was $r_0 = 14$ mm, $h = 2$ mm, $s_0 = 20$ mm, the stationary temperature was experimentally measured at 1 and 15 kHz and at voltage from 10 to 140 V.

The temperature on the surface of the piezoceramics was measured by a thermistor with a nominal resistance of 1.6 kOhm, which was welded to the inner electrode of an empty cylinder with one outlet with a reliable heat contact. The thermistor was pre-calibrated in a thermostat at 20-32°C with the help of a digital voltmeter. The calibrating error did not exceed ±0.5°C. To increase the precision and achieve the heat equilibrium between the ceramic part and the medium in the thermostat chamber, the time between the two measurements was chosen within 15-20 min. The obtained experimental data were used for the graphs in Fig. 58.

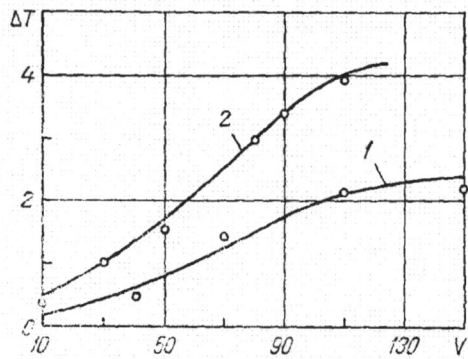

Fig. 58. Dependence of temperature T (°C) on the electric field strength V (V) and frequency (1 is f = 1 kHz; 2 is 15 kHz; the dots are experimental data)

The precision analysis of the experiment gives the following formula for the relative error, which is no more than ±8.5-9%:

$$f_{\Delta t} = \pm\sqrt{f_{\Delta R}^2 + f_{tg\alpha}^2}.$$

The experimental error when working with the voltmeter and thyristor $f_{\Delta R} = 2.5\%$ and $f_{tg\alpha} = 7.9\%$ is $f_{\Delta t} \pm 8.3\%$.

From the computer calculation for the frequency $\omega = 2\pi \cdot 10^3$ s^{-1} and $V_0 = 100$ V the maximal temperature growth is $T_{max} = 1.72\,°C$, experimentally $T_{max} = 2\,°C$. The calculations of heat emissions by formulas (7)-(9) have shown that in the case considered the heating occurs mainly due to internal mechanical losses.

The denotations in this section: s, θ are the coordinates of the points on the middle surface of the transducer shell in the meridian and circumferential directions; z is the coordinate in the radial direction; V is the electric potential; r_0 is the cylinder radius; N_s, N_θ are the meridian and circumferential tension forces in the shell; M_s, M_θ are the internal binding moments in the meridian and circumferential direction; Q_s are the cross-section cutting forces; h is the shell thickness; ρ is the density; ω is the angular frequency of oscillations; U_s, U_z are the components of the vector of displacements in the meridian direction and along the normal line to the middle surface; ε_s, ε_θ, ε_z are the relative deformations of piezoceramics; ν is the Poisson's coefficient; $E_z^{(0)}$ is the electric field strength; d_{31} is the piezomodule; s_{11}^E is the compliance of piezoceramics; C_{el} is the electric capacity; v is the volume of the transducer; E is the Young's modulus; e_{31}'', e_{33}'' are the imaginary parts of complex piezoelectric constants; $f_{\Delta R}$ is the relative error in measuring the resistance of the thyristor; $f_{tg\alpha}$ is the error of measuring the tangent to the calibration curve; α_t is the heat emission coefficient; F is the heat exchange surface; q_n is the uniformly distributed, normal to the surface of the shell, mechanical load.

12. Electrostrictive effect in FE ceramic shells at harmonic excitation

Under a strong constant electric field electrostrictive ferroelectric ceramics in the non-polar phase undergoes polarization; a considerably smaller alternating field excites operating harmonic oscillations.

The solution of the problem is the sum of functions depending on the constant $\overline{E}_3$ and variable $\widetilde{E}_3$ components of the electric field [Mason W.P. Piezoelectric crystals and their apllications to ultrasonics. N.Y.: Van Nostrend. 1950. XI].

The displacement electric field in ceramics is screened by electrons in metal electrodes. The electric induction D_3 abruptly rises within the thin layer adjacent to the electrode, and in the remaining part (along the shell thickness) $\mathbf{D}_3$ is practically constant [Chensky U.V. Sovjet Solid State Physics. 1970. Vol.12. No.2, p.586] and can be determined by the following expressions

$$\mathbf{D}_3 = \varepsilon_{33}^{T}\mathbf{E}_3 + 2\mathbf{Q}_{12}\overline{\mathbf{E}}_3(\sigma_1 + \sigma_2), \qquad (1)$$

$$\sigma_1 = \frac{1}{s_{11}^{E}(1-\mu)}[\varepsilon_1 + \mu\varepsilon_2 - 2(1+\mu)\mathbf{Q}_{12}\overline{\mathbf{E}}_3\widetilde{\mathbf{E}}_3]$$

$$\sigma_2 = \frac{1}{s_{11}^{E}(1-\mu)}[\varepsilon_2 + \mu\varepsilon_1 - 2(1+\mu)\mathbf{Q}_{12}\overline{\mathbf{E}}_3\widetilde{\mathbf{E}}_3], \qquad (2)$$

σ_1, σ_2 are the internal mechanical stresses in the meridian and circumferential directions; μ is the Poisson's coefficient; ε_{33}^{T} is the dielectric permeability; s_{11}^{E} is the compliance; Q_{12} is the electrostrictive constant;, ε_1, ε_2 are the relative strains.

Substituting (2) into (1), we obtain

$$D_3 = \varepsilon_{33}^{T}E_3 + \frac{2Q_{12}\overline{E}_3}{s_{11}^{E}(1-\mu)}(\varepsilon_1 + \varepsilon_2 - 4Q_{12}\overline{E}_3\widetilde{E}_3). \qquad (3)$$

The change of strain along the shell thickness is determined by Kirchhoff-Love hypotheses

$$\varepsilon_1 = \varepsilon_1^{(0)} + z\kappa_1; \ \varepsilon_2 = \varepsilon_2^{(0)} + z\kappa_2, \qquad (4)$$

where $\varepsilon_1^{(0)}$, $\varepsilon_2^{(0)}$ are the deformations of the middle surface of the shell in the meridian and circumferential directions; Z is the coordinate in the direction of the shell thickness; κ_1, κ_2 are the changes of the main curvatures of the middle surface.

We similarly (4) write the time-periodic component of the electric field:

$$\widetilde{E}_3 = \widetilde{E}_3^{(0)} + z\widetilde{E}_3^{(1)}e^{i\omega t}, \ \widetilde{E}_3^{(0)} = -\frac{2V_0}{h}e^{i\omega t}, \qquad (5)$$

V_0 is the electric potential; h is the shell thickness.

Substituting (4) into (2) and expressing the forces and bending moments through integrals of stresses σ_1 and σ_2, we obtain the following relations of electroelasticity

$$T_1^{H} = \varepsilon_1^{(0)} + \mu\varepsilon_2^{(0)} - 2(1+\mu)Q_{12}\overline{E}_3^{(0)}\widetilde{E}_3^{(0)},$$

$$T_2^{H} = \varepsilon_{2(0)} + \mu\varepsilon_1^{(0)} - 2(1+\mu)Q_{12}\overline{E}_3^{(0)}\widetilde{E}_3^{(0)},$$

$$M_1 = D_M[\kappa_1 + \mu\kappa_2 - 2(1+\mu)Q_{12}\overline{E}_3^{(1)}\widetilde{E}_3^{(1)}], \qquad (6)$$

$$M_2 = D_M[\kappa_2 + \mu\kappa_1 - 2(1+\mu)Q_{12}\overline{E}_3^{(1)}\widetilde{E}_3^{(2)}],$$

$$T_1^H = \frac{T_1}{D_N}, \quad T_2^H = \frac{T_2}{D_N}, \quad D_N = \frac{h}{s_{11}^E(1-\mu^2)}, \quad D_M = \frac{h^3}{12s_{11}^E(1-\mu^2)},$$

T_1, T_2 are the tensing forces in the meridian and circumferential directions; M_1, M_2 are the bending moments.

Let us substitute (4) and (5) into (3). Considering that $D_3(z) = \text{const}$, putting the coefficients at z to zero, we obtain the relations for the constant [Boriseiko V.A., Grinchenko V.T., Ulitko A.F. Sov. Appl. Mechanics. 1976. Vol.12. No.2, p.26] and variable components of the electric field:

$$\overline{E}_3 = -\frac{2\overline{V}_0}{h}, \quad \overline{E}_3^{(1)} = -\frac{K_p^2}{2d_{31}^{eq}(1-K_p^2)}(\overline{\kappa}_1 + \overline{\kappa}_2),$$

$$K_p^2 = \frac{2(d_{31}^{eq})^2}{(1-\mu)s_{11}^E\varepsilon_{33}^T}, \quad \widetilde{E}_3^{(0)} = -2\widetilde{V}_{0/h}, \quad d_{31}^{eq} = Q_{12}E_3, \qquad (7)$$

$$\widetilde{E}_3^{(1)} = \frac{16Q_{12}^2\overline{E}_3^{(0)}\overline{E}_3^{(1)}\widetilde{E}_3^{(0)} - 2Q_{12}[\overline{E}_3^{(0)}(\widetilde{\kappa}_1 + \widetilde{\kappa}_2) + \overline{E}_3^{(1)}(\widetilde{\varepsilon}_1^{(0)} + \widetilde{\varepsilon}_2^{(0)})] - \overline{E}_3^{(1)}\varepsilon_{33}^T s_{11}^E(1-\mu)}{\varepsilon_{33}^T(1-\mu)s_{11}^E - 8Q_{12}^2(\overline{E}_3^{(0)})^2}$$

In view of motion equations [Love A.E.H. A Treatise on the Mathematical Theory of Elasticity. N.Y.: Dover Publ. 1944], (6) and (7) determine the stressed state of electrostrictive ceramics. In the first place, under a constant voltage $\overline{V}_0$ the electrostatic problem is solved; the formulas, e.g. given in [Bogomol'nyi V.M. Sov. Phys. – Techn. Phys. 1981. Vol.51. No.6, p.851] use the equivalent value d_{31}^{eq} instead of the piezomodule d_{31}, as well as expressions (6) and (7).

13. Electrostrictive effect in prestressed FE ceramic shells

Under a strong constant electric field ferroelectric ceramics in the non-polar phase undergoes polarization; a considerably smaller additional field creates working displacements of the piezoelement [Mason W.P. Piezoelectric crystals and their apllications to ultrasonics. N.Y.: Van Nostrend. 1950. XI]. The electrostrictive effect helps to control the value of electromechanical coupling linearly (without hysteresis), which explains its practicality for precision devices.

Well-known papers have considered piezoelements of the simplest geometry (bars, plates) [Leung K.M., Liu K.T., Kyonka J. Large electrostrictive effect in Ba.PZT and its applications. Ferroelectrics. 1980. Vol.27. No.1/4, p.41-43]; this work aims to find relations of electroelasticity for metal-reinforced shells. Let us consider a cylindrical ferroelectric ceramic shell with a metal shell attached to it due to temperature difference.

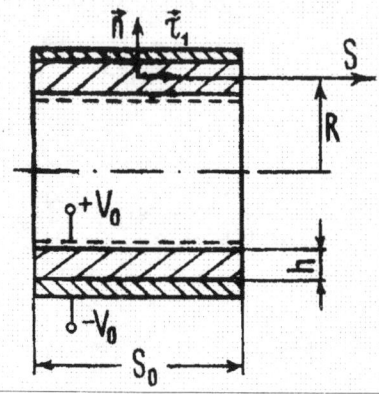

Fig. 59.

At polarization directed towards the shell thickness the equations of the electrostrictive effect related to the time-constant mechanical stresses and strains have the form

$$\varepsilon_1 = s_{11}^E \sigma_1 + s_{12}^E \sigma_2 + s_{13}^E \sigma_3 + Q_{12} E_3^2, \; \varepsilon_2 = s_{11}^E \sigma_2 + s_{12}^E \sigma_1 + s_{13}^E \sigma_3 + Q_{12} E_3^2,$$

$$\varepsilon_3 = s_{13}^E (\sigma_1 + \sigma_2) + s_{33}^E \sigma_3 + Q_{11} E_3^2, \quad (1)$$

where ε_1, ε_2, ε_3 are the relative strains in the direction of single vectors $\vec{\tau}_1$, $\vec{\tau}_2$, $\vec{n}$ (Fig. 59); σ_i ($i = 1,2,3$) are the mechanical stresses; s_{1i}^E ($i = 1,2,3$) are the elastic flexibilities of the ferroelectric ceramics; Q_{11}, Q_{12} are the electrostrictive constants; E_3 is the component of the vector of the electric field strength in the direction of the shell thickness.

Initial stresses $\overline{\sigma}_2$, $\overline{\sigma}_3$, which occur due to contact pressure between layers q_n, are determined by the methods of the flat theory of elasticity [Love A.E.H. A Treatise on the Mathematical Theory of Elasticity. N.Y.: Dover Publ. 1944]. The circular stresses σ_2 consist of two components

$\sigma_2 = \bar{\sigma}_2 + \tilde{\sigma}_2$, where $\tilde{\sigma}_2$ depend on E_3; q_n are determined in this book. The tangential contact forces that occur on the surface of ferroelectric ceramics due to temperature difference are taken into account as the surface load $q_1 = fq_n$ (f is the friction coefficient) applied to the ferroelectric ceramic shell.

The component of the vector of electric induction towards the shell thickness D_3 is expressed through the tensor of mechanical strains

$$D_3 = \varepsilon_{33}^s E_3 + e_{31}^{eq}(\varepsilon_1 + \varepsilon_2) + e_{33}^{eq}\varepsilon_3,$$

$$e_{31}^{eq} = Q_{12}E_3(c_{11}^E + c_{12}^E) + Q_{11}E_3 c_{13}^E, \quad (2)$$

$$e_{33}^{eq} = 2Q_{12}E_3 c_{13}^E + Q_{11}E_3 c_{33}^E,$$

$$\varepsilon_{33}^s = \varepsilon_{33}^T - 2Q_{12}E_3 e_{31}^{eq} - Q_{11}E_3 e_{33}^{eq}, \quad (3)$$

c_{1i} ($i = 1,2,3$) are the elastic constants; ε_{33}^s, ε_{33}^T are the dielectric permeabilities of ferroelectric ceramics, measuring correspondingly at zero strains and mechanical stresses.

Substituting expressions (3) into (2) in view of the third relation (1), we have

$$D_3 = E_3\varepsilon_{33}^T - 2Q_{12}E_3^2(c_{11}^E + c_{12}^E) + Q_{11}c_{13}^E] -$$
$$- Q_{11}E_3^2(2Q_{12}c_{13}^E + Q_{11}c_{33}^E) + E_3(\varepsilon_1 + \varepsilon_2)[Q_{12}(c_{11}^E + c_{12}^E) + Q_{11}c_{13}^E] +$$
$$+ E_3(2Q_{12}c_{13}^E + Q_{11}c_{33}^E)[s_{13}^E(\sigma_1 + \sigma_2) + s_{33}^E\tilde{\sigma}_3 + Q_{11}E_3^2] \quad (4)$$

(From here on the sign above $\tilde{\sigma}_1$ and $\tilde{\sigma}_2$, which depend on E_3 and q_n, will not be used for the sake of simpler notification).

In the case under consideration the preliminary mechanical stresses $\bar{\sigma}_3$ (which occur because of contact pressure q_n) are calculated by the formula

$$\bar{\sigma}_3 = -q_n\left(\frac{1}{2} + \frac{z}{h}\right),$$

Z is the coordinate in the direction of the normal line to the middle surface of the shell with a thickness h (Fig. 59).

The first two equations (1) give

$$\sigma_1 = \frac{1}{s_{11}^E(1-\mu^2)}(\varepsilon_1 + \mu\varepsilon_2 - \hat{E}_3)$$

$$\sigma_2 = \frac{1}{s_{11}^E(1-\mu^2)}(\varepsilon_2 + \mu\varepsilon_1 - \hat{E}_3) \tag{5}$$

$$\mu = -\frac{s_{12}^E}{s_{11}^E}, \quad \hat{E}_3 = (1+\mu)(Q_{12}E_3^2 - q_n s_{13}^E/2).$$

The distribution of meridian and circumferential deformations along the shell thickness is determined by Kirchhoff-Love hypotheses

$$\varepsilon_1 = \varepsilon_1^{(0)} + z\kappa_1, \quad \varepsilon_2 = \varepsilon_2^{(0)} + z\kappa_2, \tag{6}$$

$\varepsilon_1^{(0)}$, $\varepsilon_2^{(0)}$ are the strains of the middle surface of the shell; κ_1, κ_2 are the changes of its main curvatures.

The electric field strength can be measured by similar (6) expressions [Drumheller D.S., Kalnins A. Dynamic shell theory for ferroelectric ceramic. JASA. 1970. Vol. 47. No.5, p.1343-1353]

$$E_3 = E_3^{(0)} + zE_3^{(1)}, \quad E_3^{(0)} = -\frac{2V_0}{h}, \tag{7}$$

where $\pm V_0$ are the electric potentials on the electrodes situated on the surfaces of the shell with coordinates $z = \pm h/2$.

The value D_3 undergoes practically no changes along the shell thickness [Chensky E.V. On mono-domain polarization of ferroelectrics in phase transfer of the first kind. Sov. Phys. Solid State. 1970. Vol.12. No.3, p.586-591]. Following [Boriseiko V.A., Grinchenko V.T., Ulitko L.F. Electroelasticity relations for piezoceramic rotation shells. Sov. Appl. Mechanics. 1976. Vol.12. No.2, p.26-33], we substitute (5)-(7) into (4); in the expression obtained for D_3 putting the total of the components at z to zero, we determine $E_3^{(1)}$

$$E_3^{(1)} = \frac{F_3^{(0)}[hA(\kappa_1,\kappa_2)+(E_3^{(0)})^2 q_n s_{33}^E Q_{11}(2Q_{12}c_{13}^E + Q_{11}c_{33}^E)]}{h[B(\varepsilon_1^{(0)},\varepsilon_2^{(0)})+(F_3^{(0)})^2 c(q_n)]}, \tag{8}$$

$$A(\kappa_1,\kappa_2 = (\kappa_1 + \kappa_2)\times$$

$$\times\left[Q_{12}(c_{11}^E + c_{12}^E)+Q_{11}c_{13}^E + \frac{s_{13}^E(2Q_{12}c_{13}^E + Q_{11}c_{33}^E)}{s_{11}^E(1-\mu)}\right],$$

$$B(\varepsilon_1^{(0)}, \varepsilon_2^{(0)}) = \varepsilon_{33}^T + (\varepsilon_1^{(0)} + \varepsilon_2^{(0)})[Q_{12}(c_{11}^E + c_{12}^E) + Q_{11}c_{13}^E] +$$

$$+ \frac{s_{13}^E}{s_{11}^E(1-\mu)}(2Q_{12}c_{13}^E + Q_{11}c_{33}^E)(\varepsilon_1^{(0)} + \varepsilon_2^{(0)} + q_n s_{13}^E),$$

$$c(q_n) = -6Q_{12}^2(c_{11}^E + c_{12}^\#) - 8Q_{11}Q_{12}c_{13}^E - 3Q_{11}c_{33}^E -$$

$$- 4Q_{12}c_{13}^E - \frac{6s_{13}^E Q_{12}}{s_{11}^E(1-\mu)}(2Q_{12}c_{13}^E + Q_{11}c_{33}^E) +$$

$$+ 3s_{33}^E q_n\left(Q_{11}Q_{12}c_{13}^E + \frac{Q_{11}^2 c_{33}^E}{2}\right).$$

Using relations (6), (7) and expressing forces T_1, T_2, moments M_1, M_2 through integrals of stresses, we obtain the following relations:

$$T_1 = D_\Lambda(\varepsilon_1^{(0)} + \mu\varepsilon_2^{(0)} - \overset{\Lambda}{E}_3^{(0)}), \quad T_2 = D_\Lambda(\varepsilon_2^{(0)} + \mu\varepsilon_1^{(0)} - \overset{\Lambda}{E}_3^{(0)}), \qquad (9)$$

$$M_1 = D_M(\kappa_1 + \mu\kappa_2 - \overset{\Lambda}{E}_3^{(1)}), \quad M_2 = D_M(\kappa_2 + \mu\kappa_1 - \overset{\Lambda}{E}_3^{(1)}),$$

$$D_N = \frac{h}{s_{11}^E(-\mu^2)}, \quad \overset{\Lambda}{E}_3^{(0)} = (1+\mu)\left[Q_{12}(E_3^{(0)})^2 - \frac{q_n s_{13}^E}{2}\right],$$

$$D_M = \frac{h^3}{12s_{11}^E(1-\mu^2)}, \quad \overset{\Lambda}{E}_3^{(1)} = (1+\mu)\left(2Q_{12}E_3^{(0)}E_3^{(1)} - \frac{s_{13}^E q_n}{h}\right).$$

At axisymmetrical deformations expression (8) may be simplified, if $\varepsilon_1^{(0)}$ and $\varepsilon_2^{(0)}$ are determined by the momentless theory from the following equilibrium equations, written in the dimensionless form [Armenakas A.E. Effect of initial strains on the oscillations of free-mounted cylindrical shells. AIAA Journ. 1964. Vol.2. No.9, p.115-122]:

$$\frac{dT_1^H}{ds} - \overline{T}_2^H\frac{dW^H}{d\tilde{s}} - q_1 = 0, \quad \frac{dN_1^H}{ds} - T_2^H - q = 0,$$

$$-\overline{T}_{2(1)}^H\frac{dW^H}{ds} = N_1^H, \quad (10)$$

$$T_1^H = \frac{T_1}{D_N}, \quad T_2^H = \frac{T_2}{D_N}, \quad \overline{T}_{2(1)}^H = \frac{h\overline{T}_2^H}{2R}, \quad \overline{T}_2^H = \frac{\overline{T}_2}{D_N}, \quad q_1 = \frac{fq_n R}{D_N},$$

$$q = \frac{q_n R}{D_N}, \quad \tilde{s} = \frac{s}{R},$$

S is the linear coordinate directed along the generating shell, R is the radius of its middle surface, $\overline{T}_2 = -q_n R(1 + h/2R)$ is the initial circular compression force, W^H is the component of the vector by which the middle surface of the shell is displaced towards the single vector $\bar{n}$ (Fig. 59) ($W^H = W/R$).

Let us express deformations through displacements

$$\varepsilon_1^{(0)} = \frac{dU^H}{ds}, \ \varepsilon_2^{(0)} = W^H, \ \kappa_1 = -\frac{1}{R}\frac{d^2 W^H}{ds^2}, \ \kappa_2 = 0. \quad (11)$$

Using (7) (where $E_3 \approx E_3^{(0)}$), from (10), (11) we obtain an equation for determining $\varepsilon_1^{(0)}$ and $\varepsilon_2^{(0)}$

$$\frac{d^2 W^H}{ds^2} + r^2 W^H = f_0 \quad (12)$$

$$r^2 = \frac{1 + \mu \overline{T}_2^H - v^2}{\overline{T}_{2(1)}^H}, \ f_0 = \frac{1}{\overline{T}_{2(1)}^H}\left[(1-\mu)\hat{E}_3^{(0)} - \int_0^{\tilde{s}_0/2} q_1 d\tilde{s} - q\right].$$

From the following edge conditions we find the solution of equation (12), and then $\varepsilon_1^{(0)}$ and $\varepsilon_2^{(0)}$

$$\left.\frac{dW^H}{ds}\right|_{\tilde{s}=0} = 0, \ \left.T_1^H\right|_{\tilde{s}=\tilde{s}_0/2} = 0 \qquad (13)$$

$$\varepsilon_1^{(0)} = (\overline{T}_2^H - \mu)\left[\left(\frac{\hat{E}_3^{(0)}}{1-\mu} - \frac{q_n}{1-\mu^2}\right)\frac{\cos r\tilde{s}}{\cos(r\tilde{s}_0/2)} - \frac{f_0 \cos r\tilde{s}}{r^2 \cos(r\tilde{s}_0/2)}\right] +$$

$$+ \hat{E}_3^{(0)} + \frac{q_1 \tilde{s}_0}{2} \quad (14)$$

$$\varepsilon_2^{(0)} = \frac{1}{(1-\mu^2)}[(1+\mu)\hat{E}_3^{(0)} - q]\frac{\cos r\tilde{s}}{\cos(r\tilde{s}_0/2)} - \frac{f_0 \cos r\tilde{s}}{r^2 \cos(r\tilde{s}_0/2)} \quad (15)$$

Substituting (15), (14) into (8), we obtain an expression for $E_3^{(1)}$ dependent only on one function κ_1 ($\kappa_2 = 0$), which gives an explicit solution. For this we use a complete system of equations (10), in which the third equation contains $M_1^H = M_1/RD_N$ [Armenakas A.E. Effect of initial strains on the oscillations of free-mounted cylindrical shells. AIAA Journ. 1964. Vol.2. No.9, p.115-122]

$$\frac{dM_1}{d\tilde{s}} - \overline{T}_{2(1)}^H \frac{dW^H}{d\tilde{s}} = N_1^H. \quad (16)$$

The system of equations (9)-(11) and (16) is reduced to one equation

$$\frac{d^4 W^H}{d\tilde{s}^4} - 2r^2 \frac{d^2 W^H}{d\tilde{s}^2} + s_1^4 W^H = f_1, \quad 2r^2 = \frac{\overline{T}_{2(1)}^H D_N R^2}{D_M}, \quad (17)$$

$$s_1^4 = \frac{1 + \mu \overline{T}_2^H - \mu^2}{m}, \quad m = \frac{D}{D_M R^2}, \quad f_1 = -\frac{\mu}{m} \int_0^{s_0/2} q_1 d\tilde{s}.$$

The solution of equation (17) is found from the analysis of the roots of its characteristic equation in view of the parity of the function W^H regarding the origin of coordinates $\tilde{s} = 0$ (located in the centre of the shell). At $s_1 > r$ it has the form

$$W^H = C_1 \text{ch}\alpha\tilde{s} \cos\beta\tilde{s} + C_2 \text{sh}\alpha\tilde{s} \sin\beta\tilde{s} + f_1 / s_1^4, \quad (18)$$

$$\alpha = \left(\frac{s_1^2 + r^2}{2}\right)^{1/2}, \quad \beta = \left(\frac{s_1^2 - r^2}{2}\right)^{1/2},$$

where α, β are real positive numbers.

The integrating constants C_1, C_2 are determined from edge conditions (13), where the second edge condition corresponds at $\tilde{s} = \tilde{s}_0 / 2$ to the following expression

$$m_1 \frac{d^4 W^4}{d\tilde{s}^4} + a \frac{d^2 W^H}{d\tilde{s}^2} + bW^H + C = 0,$$

$$m_1 = -\frac{m}{\mu}, \quad a = \frac{-\overline{T}_{2(1)}^H}{\mu}, \quad b = \frac{\mu^2 - 1}{\mu}, \quad C = \frac{(1-\mu)\hat{E}_3^{(0)} - q}{\mu}.$$

The relations of electroelasticity (9) are obtained for thin shells of arbitrary shape. In the case of the cylindrical shell they can be simplified, and the solutions of (18) will be determined by A.N.Krylov's trigonometric functions, given in various reference sources.

14. Amplitude-frequency characteristics of a linear piezomotor

To increase the oscillation amplitude of a piezoelement, it is preferable to use piezoceramic rings that are open in the circumferential direction. A layout of such a PM is given in Fig. 65. The construction consists of two thickness-polarized ceramic half-rings with a moveable metal plate between them. One half-ring is rigidly mounted, the other is clamped by an adjusting screw with a spring. Only one ring, on which the moved object is mounted with an elastic contact, can be used as a linear drive. At electric excitation the half-ring moves against the immobile foundation.

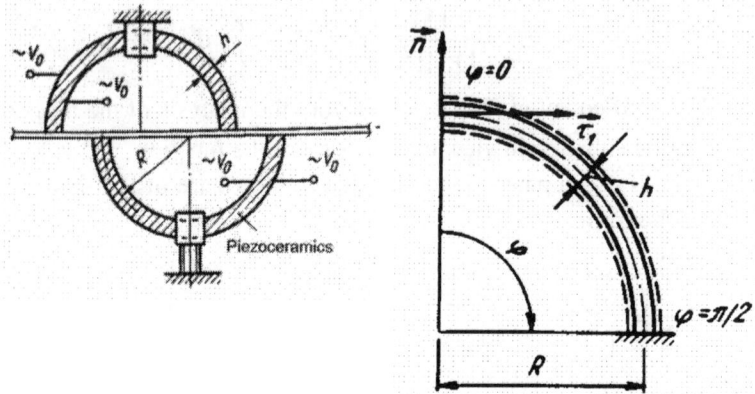

Fig. 65. Piezoelectric motor of linear displacements

Fig. 66. Analytical model of a piezo-element

Below we calculate a fourth of the console-mounted piezoceramic ring (Fig. 66). Numerical calculation determines the displacement vector and the trajectory of the free end movement.

We consider a bar piezoceramic element with an axis represented by a circle arc with the radius R, limited by coordinates $0 < \varphi < \pi/2$, with electric potentials $V = \pm V_0 e^{i\omega t}$ applied to its surface $z = \pm h/2$ (h is the height of the rectangular cross section; Fig. 66 shows electrodes by the dashed line). The edge $\varphi = \pi/2$ is rigidly mounted, whereas $\varphi = 0$ is free from external forces.

The dimensionless movement equations of the curvilinear piezoceramic bar have the form

$$\frac{dT_2^N}{d\varphi} + Q_2 = -\lambda v^N, \quad Q = \frac{dM_2^N}{d\varphi},$$

$$\frac{dQ_2^N}{d\varphi} - T_2^N = -\lambda w^N,$$

$$\left. T_2^N = \frac{T_2}{D_N}, Q_2^N = \frac{Q_2}{D_n}, \lambda = \frac{R^2 h \rho \omega^2}{D_n}, \right\} \quad (1)$$

$$D_N = \frac{h}{s_{11}^E(1-\mu^2)}, \quad M_2^N = \frac{M_2}{RD_N}, \quad w^N = \frac{w}{R}, \quad v^N = \frac{v}{R},$$

where T_2 are the tension forces; Q_2 are the cross-section intersecting forces; M_2^N are the bending moments; w, v are the components of the displacement vector in the direction of the unit vectors n, τ (see Fig. 66); ρ is the density; s_{11}^E is the compliance of piezoceramics; μ is the Poisson's coefficient; φ is the angular coordinate.

In case of thickness polarization of the bar, the electroelasticity relations have the form

$$T_2^{N'} = (1-\mu)[(1+\mu)\varepsilon_\varphi - E_0], \quad M_2 = D_M \kappa_\varphi,$$

$$D_M = \frac{h^3\gamma}{12s_{11}^E(1-\mu^2)}, \quad E_0 = (1+\mu)d_{31}E_z^0, \quad E_z^0 = \frac{2V_0}{h},$$

where ε_φ is the relative strain of the bar axis; E_z^0 is the component of the electric field strength vector in the direction of thickness; d_{31} is the piezomodule; κ_φ is the change of the bar axis curvature.

$$\gamma = \left[1 + \frac{(1+\mu)}{2}\frac{\kappa_p^2}{(1-\kappa_p^2)}\right]$$

$$R_p^2 = \frac{2}{(1-\mu)}\frac{d_{31}^2}{s_{11}^E \varepsilon_{33}^T},$$

where ε_{33}^T is the dielectric permeability of ceramics.

Let us express the strains through displacements:

$$\varepsilon_s = E_0 - \mu\varepsilon_\varphi, \quad \varepsilon_\varphi = w^N + \frac{dv^N}{d\varphi},$$

$$\kappa_\varphi = -\frac{1}{R}\left(\frac{d^2w}{d\varphi^2} - \frac{dw}{d\varphi}\right), \quad (3)$$

ε_s are the strains in the direction of the unit vector $\vec{\tau}_2$, which is perpendicular to vectors $\vec{n}$ and $\vec{\tau}_1$.

In view of (3) equations (2) will have the form

$$T_2^N = (1-\mu)[(1+\mu)(w^N + dv^N/d\varphi) - E_0), \quad (4)$$

$$M_2^N = \frac{h^2\gamma}{12R^2}\left(-\frac{d^2w^N}{d\varphi^2} + \frac{dv^N}{d\varphi}\right). \quad (5)$$

Edge conditions for the main equation system (1), (4), (5) will be:

$$M_2^N = T_2^N = 0 \text{ at } \varphi = 0,$$

$$v^N = w^N = \frac{dw^N}{d\varphi} = 0 \text{ at } \varphi = \frac{\pi}{2}. \quad (6)$$

Let us present the equation system (1), (4), (5) as a matrix:

$$\dot{\vec{X}} = A\vec{x} + \vec{F}_N,$$

$$
\dot{\vec{x}} = \begin{vmatrix} \dfrac{dT_2^N}{d\varphi} \\ \dfrac{dQ_2^N}{d\varphi} \\ \dfrac{dM_2^N}{d\varphi} \\ \dfrac{dv^N}{d\varphi} \\ \dfrac{dw^N}{d\varphi} \end{vmatrix}
\qquad
\vec{x} = \begin{vmatrix} T_2^N \\ Q_2^N \\ M_2^N \\ v^N \\ \dot{w}^N \end{vmatrix}
\qquad
F = \begin{vmatrix} 0 \\ 0 \\ 0 \\ E_0 \\ \dfrac{E_0}{1+\mu} \\ \dfrac{E_0}{1+\mu} \end{vmatrix},
$$

where $dw^N/d\varphi = \dot{w}^N$, A is the matrix of coefficients with unknown functions.

Opening the determinant of the equation $\det(A - \beta E)$, where E is the diagonal matrix, we find the roots of the characteristic equation:

$$\beta^6 + K_1\beta^4 + K_2\beta^2 + K_3 = 0,$$

$$K_1 = 2 + \frac{\lambda}{1-\mu^2}, \quad K_2 = 1 - \frac{\lambda}{1-\mu^2} - \lambda\psi,$$

$$K_3 = \psi\lambda\left(1 - \frac{\lambda}{1-\mu^2}\right), \quad \psi = \frac{12R^2}{h^2\gamma}.$$

From the matrix A and the vector of eigenvalues β_i ($i = 1,2,3,...,6$) the solution of the main equation system will be

$$\vec{x} = sCe^{\beta\varphi}T - A^{-1}F,$$

where S is the matrix of eigenvectors; C is the diagonal matrix relative to the integration constants, determined from the edge conditions; I is the unit matrix (one column with units).

This algorithm was realized on the computer. The displacements of the piezoceramic ring obtained by the calculation of a piezoceramic PZT-4 bar ($R = 20$ mm, $h = 2$ mm, $V_0 = 100$ V and $f = 1$ kHz), are shown in Fig. 67.

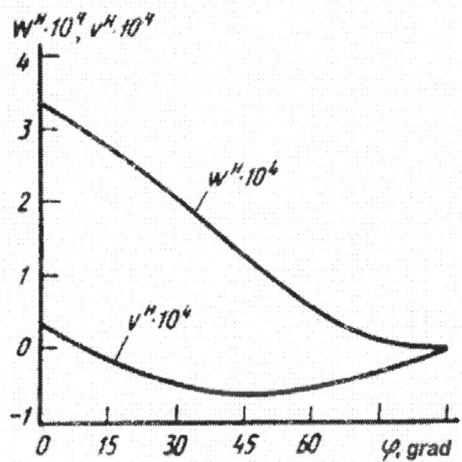

Fig. 67. Displacement graphs of the PZ ring-like element

To determine the displacements of the free end of the bar and the first lowest resonance frequency, we use the momentless theory.

From the third equation (1) we have $T_2^N = -\lambda w^N$. At $M_2^N = 0$ it follows from equation (5)

$$\frac{d^2 w^N}{d\varphi^2} = \frac{dv^N}{d\varphi}. \qquad (7)$$

Equating the right-hand part of equation (4) to the value $-\lambda w^N$ in view of (6), we obtain

$$\frac{d^2 w^N}{d\varphi^2} + n^2 w^N = \frac{E_0}{(1-\mu^2)}, \qquad (8)$$

$$n^2 = \frac{1-\mu^2-\lambda}{1-\mu^2}.$$

The solution of equation (8) has the form

$w^N = \overline{A}\sin n\varphi + \overline{B}\cos n\varphi + E_0/(1+\mu)n^2$.

The integration constants $\overline{A}$ and $\overline{B}$ are determined from the edge conditions (6).

From conditions $w^N(\varphi = 0) \to \infty$ (at $\varphi = 0$) we find the lowest angular frequency of resonance oscillations: $\omega^* = (R^2\rho s_{11}^E)^{-1/2}$. For the considered example $\omega^* = 1.14 \cdot 10^5$ s-1, as calculated on the computer by the algorithm given above $\omega^* = 1.14 \cdot 10^5$ s-1.

As follows from the graphs in Fig. 67 and the calculations (in the frequency range from 1 to 30 kHz), at frequencies lower than the first resonance frequency f = 23 kHz the displacements of W are approximately by one order greater than V (directed tangentially towards the neutral axis of the bar), i.e. the end of the bar (the contact point of piezoceramics with a displaced object) moves along the elliptic trajectory with the relations of half-axes within (4:1) – (10:1). Hence, piezoceramic elements can be used as a motor of linear and angular displacements (the ratio w / V changes with the growth of electric excitation frequency, at F = 30 kHz w / V= 2.4).

The displacements of the free end of the piezoceramic bar (with the co-ordinate $\varphi = 0$) at the stated above parameters and excitation conditions depending on the excitation frequency are given in Tables 9 and 10.

Dis-place-ments	f, kHz				
	1	10	20	21	22
v^N	3.08·10-5	-13.02·10-5	-1.06·10-4	-1.64·10-4	-4.48·10-4
w^N	3.41·10-5	-9.77·10-5	15.89·10-3	103.12·10-3	130.04·10-3

Table 9. Components of the displacement vector of the free end of the bar

Displace-ments	f, kHz			
	23	24	25	30
v^N	5.12·10-4	1.6·10-4	9.75·10-4	4,05·10-4
w^N	2470·10-4	321·10-4	147·10-4	-9.77·10-4

Table 10. Components of the displacement vector of the free end of the bar

As follows from Table 10, the reversibility of the piezoceramic ring motor can be ensured by changing the frequency of electric excitation (or with the help of separate electrodes).

15. Calculation of piezoceramic transducers with uneven thickness polarization

Let us consider a piezoceramic shell of revolution, which is pre-polarized under a strong electric field and is free from external mechanical loads. Its outer surfaces with coordinates $z = \pm h/2$ (h is the shell thickness) are supplied with electric potentials $V = \pm V_0$. The equations of the inverse piezoeffect in this case have the form [Drumheller D.S., Kalnins A. JASA. 1979. Vol. 47. No.5, p.1343; Boriseiko V.A., Grinchenko V.T., Ulitko L.F. Electroelasticity relations for piezoceramic rotation shells. Sov. Appl. Mechanics. 1976. Vol.12. No.2, p.26-33]:

$$\left.\begin{aligned}
\varepsilon_1 &= S_{11}^E \sigma_1 + S_{12}^E \sigma_2 + d_{31} E_3, \\
\varepsilon_2 &= S_{11}^E \sigma_2 + S_{12}^E \sigma_1 + d_{31} E_3, \\
D_3 &= \varepsilon_{33}^T E_3 + d_{31}(\sigma_1 + \sigma_2),
\end{aligned}\right\} \tag{1}$$

where ε_i ($i = 1,2$) is the relative strain of the middle surface of the shell in the axis and circumferential directions; S_{1i}^E are the elastic flexibilities of ceramics; σ_i are the mechanical stresses; d_{31} is the piezomodule; E_3 is the electric field strength; D_3 is the electric induction; ε_{33}^T is the dielectric permeability.

When deducing the relations determining the mechanical stresses in regard to the strains and electric field strength, we suppose the following:
- the distribution of the electric field strength along the thickness of the piezoceramic shell corresponds to Kirchhoff-Love hypotheses for elastic strains [Love A.E.H. A Treatise on the Mathematical Theory of Elasticity. N.Y.: Dover Publ. 1944; Drumheller D.S., Kalnins A. JASA. 1979. Vol. 47. No.5, p.1343; [Boriseiko V.A., Grinchenko V.T., Ulitko L.F. Electroelasticity relations for piezoceramic rotation shells. Sov. Appl. Mechanics. 1976. Vol.12. No.2, p.26-33]

$$E_3 = E_3^{(0)} + z E_3^{(1)}, \quad E_3^{(0)} = -2V_0/h, \tag{2}$$

where E_3^0 is the pre-polarization of ceramics; $E_3^{(1)}$ is the constant value not depending on Z;
- electric induction and piezoelectric modules linearly depend on the co-ordinate Z (Fig. 68).

- in line with experimental data the dielectric permeability ε_{33}^T is considered constant along the thickness of the piezoceramic shell.

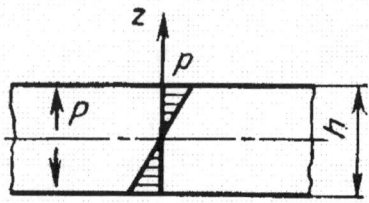

Fig.68. Distribution of polarization along the thickness of a piezoelectric transducer

The vector of electric induction $\vec{D}_3$ consists of two components which differ in their physics. One is determined by the external electric field $\vec{E}_3$, the other, polarization $\vec{P}$, depends on the eigen electrophysical properties of the polar DE

$$\vec{D}_3 = \varepsilon_{33}^T \vec{E}_3 + \vec{P} \ .(3)$$

In accordance with (2) and (3) let us express the change of induction along thickness as [Drumheller D.S., Kalnins A. JASA. 1979. Vol. 47. No.5, p.1343]

$$D_3 = D_3^{(0)} + z D_3^{(1)}, \qquad (4)$$

where D_3^0, $D_3^{(1)}$ are the constant values which do not depend on z.

For the most common piezoceramic materials like PZT-4, when relative polarization changes from 2 to 10, piezomodules increase in direct proportion by 2.2-5 times [Land C.E., Smith G.W., Westgate C.R. IEEE Trans. Sonics and Ultrasonics. 1964. Vol.SU-1. No. 1, p.8; Berlincourt D. JASA. 1964. Vol.36. No.8, p.515]. Dielectric permeability changes no more than by 7-8% and is therefore considered constant along the shell thickness

According to the experimental data, we take

$$d_{31} = d_{31}^{(0)} + z d_{31}^{(1)}, \ (d_{31}^{(0)}, \ d_{31}^{(1)} = const \ ). \qquad (5)$$

From the first two equations of the main system (1) let us express the mechanical stresses σ_1 and σ_2 through other values:

$$\sigma_1 = [S_{11}^E(1-\mu^2)]^{-1}[\varepsilon_1 + \mu\varepsilon_2 - (1+\mu)d_{31}E_3],$$
$$\sigma_2 = [S_{11}^E(1-\mu^2)]^{-1}[\varepsilon_2 + \mu\varepsilon_1 - (1+\mu)d_{31}E_3], \tag{6}$$

where $\mu = -S_{12}^E / S_{11}^E$ is the Poisson's coefficient.

Strains are expressed through displacements in accordance with Kirchhoff-Love hypotheses [Love A.E.H. A Treatise on the Mathematical Theory of Elasticity. N.Y.: Dover Publ. 1944]:

$$\varepsilon_1 = \varepsilon_1^{(0)} + z\kappa_1, \quad \varepsilon_2 = \varepsilon_2^{(0)} + z\kappa_2, \tag{7}$$

where $\varepsilon_1^{(0)}$, $\varepsilon_2^{(0)}$, κ_1, κ_2 are the strains and changes of the main curvatures of the middle shell surface.

Substituting (2), (4), (5) and (7) in the last equation of system (1), we obtain the equality

$$D_3^{(0)} + zD_3^{(1)} = \varepsilon_{33}^T(1-R_p^2)E_3^{(0)} + \frac{d_{31}^{(0)}(\varepsilon_1^{(0)} + \varepsilon_2^{(0)})}{S_{11}^E(1-\mu)} +$$

$$+ z\left[\frac{d_{31}^{(1)}(\varepsilon_1^{(0)} + \varepsilon_2^{(0)})}{S_{11}^E(1-\mu)} + \varepsilon_{33}^T(1-R_p^2)E_3^{(1)} + \frac{d_{31}^{(0)}(\kappa_1 + \kappa_2)}{S_{11}^E(1-\mu)}\right] +$$

$$+ z^2\frac{d_{31}^{(1)}(\kappa_1 + \kappa_2)}{S_{11}^E(1-\mu)}, \tag{8}$$

$$R_p^2 = \frac{2}{(1-\mu)}\frac{d_{31}^2}{S_{11}^E\varepsilon_{33}^T}. \tag{9}$$

In (9) d_{31}^2 is negligible (for PZT-4 $d_{31} = -10^{-10}$ C/N), hence with practical precision we take d_{31} in (9) as an averaged constant value.

Taking the equalities in the left-hand part (8) $D_3^{(1)} = \varepsilon_{33}^T E_3^{(1)}$ and equating the coefficients at Z in its left- and right-hand parts, we find

$$E_3^{(1)} = \frac{d_{31}^{(0)}(\kappa_1 + \kappa_2) + d_{31}^{(1)}(\varepsilon_1^{(0)} + \varepsilon_2^{(0)})}{R_p^2\varepsilon_{33}^T S_{11}^E(1-\mu)}. \tag{10}$$

Substituting (2), (5), (7) and (10) into (6) and expressing internal mechanical forces T_1, T_2 and moments M_1, M_2 through integrals of σ_1 and σ_2 by the shell thickness, we obtain the following electroelasticity relations:

$$T_1 = D_T[\varepsilon_1^{(0)} + \mu\varepsilon_2^{(0)} - (1+\mu)d_{31}^{(0)}E_3^{(0)} + \overline{A}],$$

$$T_2 = D_T[\varepsilon_2^{(0)} + \mu\varepsilon_1^{(0)} - (1+\mu)d_{31}^{(0)}E_3^{(0)} + \overline{A}],$$

$$M_1 = D_M[\kappa_1 + \mu\kappa_2 - (1+\mu)d_{31}^{(1)}E_3^{(0)} + A_1],$$

$$M_2 = D_M[\kappa_2 + \mu\kappa_1 - (1+\mu)d_{31}^{(1)}E_3^{(0)} + A_1], \qquad (11)$$

where

$$\overline{A} = -\frac{h^2(1+\mu)d_{31}^{(1)}}{12\overline{\varepsilon}_{33}^T}A,$$

$$A_1 = \frac{(1+\mu)d_{31}^{(0)}}{\overline{\varepsilon}_{33}^T}A,$$

$$A = \frac{d_{31}^{(0)}(\kappa_1 + \kappa_2) + d_{31}^{(1)}(\varepsilon_1^{(0)} + \varepsilon_2^{(0)})}{S_{11}^E(1-\mu)},$$

$$D_T = h/\overline{S}_{11}^E; \quad D_M = h^3/(12\overline{S}_{11}^E),$$

$$\overline{S}_{11}^E = S_{11}^E(1-\mu); \quad \overline{\varepsilon}_{33}^T = \varepsilon_{33}^T R_p^2.$$

To illustrate the use of formulas (11), let us consider a piezoelectric actuator in the shape of a console-mounted piezoceramic plate with a length $\ell = 35$ mm, width 5 and thickness 0.35 mm, free from external mechanical loads. The bend of the free end of the plate under an external electric field is determined from the following conditions: $T_1 = M_1 = 0$ and $\kappa_1 = $ const along the plate length ($T_2 = M_2 = \varepsilon_2^{(0)} = \kappa_2 = 0$). It follows from the first and third equations (11)

$$\varepsilon_1^{(0)} - (1+\mu)d_{31}^{(0)}E_3^{(0)} - \frac{h^2}{12}(1+\mu)\frac{d_{31}^{(1)}(d_{31}^{(0)}\kappa_1 + d_1^{(1)}\varepsilon_1^{(0)})}{S_{11}^E(1-\mu)} = 0 \quad (12)$$

$$\kappa_1 - (1+\mu)d_{31}^{(1)}E_3^{(0)} - \frac{(1+\mu)d_{31}^{(0)}(d_{31}^{(0)}\kappa_1 + d_{31}^{(1)}\varepsilon_1^{(0)})}{\overline{\varepsilon}_{33}^T S_{11}^E(1-\mu)} = 0.$$

When the polarization field strength E_p is much larger than the working field strength E_3, we take $d_{31}^{(0)} = 0$ to a first approximation. It follows from (12) that $\varepsilon_1^{(0)} = 0$, whereas $\kappa_1 = (1+\mu)d_{31}^{(1)}E_3^{(0)}$. The bend of the free end of the console piezoelectric plate is expressed by the formula $f = \ell^2\kappa_1/2$ [Timoshenko S.P., Lessels J.M., Mech J.M. Applied Elasticity. East Pittsburg: Westinghouse Techn. Night School Press, 1925]. For a piezoceramic PZT-4 actuator at $E_3^{(0)} = 50$ V/mm and the geometry stated above, the bend $f = 5.6$ μm.

If the working field strength $E_3^{(0)}$ is comparable to the polarization field strength E_p , the piezomodule $d_{31}^{(0)}$ can be found from the relation $d_{31}^{(0)} / d_{31}^{(1)} = E_3^{(0)} / E_p$, in view of which we obtain $\varepsilon_{(1)}^0$ from (12), and κ_1 from (13). We calculate the bend f by the formula given.

16. Calculation of piezoelectric IR modulators

Piezoelectric plates, unevenly polarized along thickness, are considered in [Williams W.S. Ferroelectrics. 1983. Vol. 51. No.1/2, p. 61; Kunkel H.A., Auld B.A. Ibid. P. 99; Lang S.B., Yin-Qing-Rui. Ferroelectrics. 1987. Vol.76, p.449; Bogomol'nyi V.M. Measurement Techniques. 1992. No.9, p.39; Rozenson A.E., Yakovenko N.A. Sov. Phys. – Techn. Phys. 1994. Vol.64. No.7, p.191]. Experimental data on determining uneven polarization along thickness in piezoceramic samples are given in [Lang S.B., Yin-Qing-Rui. Ferroelectrics. 1987. Vol.76, p.449]. This paragraph contains formulas for calculating mechanical stresses in piezoceramic plates and shells depending on strains and electric field strengths, as well as a method for calculating the bend of a console piezoceramic plate.

We consider a piezoceramic shell in which a strong constant electric field causes polarization, and a considerably smaller field creates working displacements of the piezoelectric transducer. In service the outer surfaces of the pre-polarised piezoceramic shell with coordinates $z = \pm h/2$ (h is the shell thickness) were supplied with electric potentials $V = \pm V_0 e^{i\omega t}$ (ω is the circular or cyclic frequency, t is the time).

When deducing the electroelasticity relations of the linear inverse piezoeffect, we suppose the following:

1. Pre-polarization P (z), piezomodules and electric induction linearly depend on the coordinate z, marked off the middle surface of the shell, and are determined by expressions

$$P = P_0 \left(\frac{1}{2} + \frac{z}{h} \right), (1)$$

$$\mathbf{d}_{31} = \mathbf{d}_{31}^{(0)} \left(\frac{1}{2} + \frac{\mathbf{z}}{\mathbf{h}} \right) \quad \mathbf{D} = \mathbf{D}_3^{(0)} \left(\frac{1}{2} + \frac{\mathbf{z}}{\mathbf{h}} \right), \quad (2)$$

where P_0, d_{31}^0, D_3^0 are accordingly polarization, piezomodule and electric induction near the outer surface of the shell with the coordinate $z = h/2$ (Fig. 69).

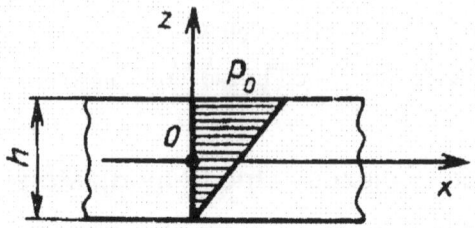

Fig.69. Distribution of polarization along the thickness of a piezoceramic plate

2. The distribution of the electric field strength along the thickness of the piezoceramic shell $E_3(z)$ corresponds to Kirchhoff-Love hypotheses for bending elastic strains [Love A.E.H. A Treatise on the Mathematical Theory of Elasticity. N.Y.: Dover Publ. 1944; Drumheller D.S., Kalnins A. *JASA*. 1970. Vol.47. No.5, p.1343; Boriseiko V.A., Grinchenko V.T., Ulitko L.F. Electroelasticity relations for piezoceramic rotation shells. *Sov. Appl. Mechanics*. 1976. Vol.12. No.2, p.26-33]. The physical essence of this supposition is that in the common case at the inverse piezoeffect the electric field strength consists of two components: the constant, which corresponds to the tension of the middle surface of the shell, and the variable, which linearly depends on the coordinate z corresponding to the bend of the middle surface of the shell:

$$E_3 = E_3^{(0)} + zE_3^{(1)}, \; E_3^{(0)} = -\frac{2V_0}{h}, \; E_3^{(1)} = \text{const}. (3)$$

3. In accordance with experimental data [Land C.E., Smith G.W., Westgate C.R. *IEEE Trans. Sonics and Ultrasonics*. 1964. Vol.SU-11. No.1, p.8; Berlincourt D. *JASA*. 1964. Vol.36. No.8, p.515] the dielectric permeability ε_{33}^T is considered an averaged value constant along the thickness of the piezoceramic shell.

The equations of the inverse piezoeffect for a piezoceramic shell without external mechanical loads have the form

$$\varepsilon_1 = S_{11}^E \sigma_1 + S_{12}^E \sigma_2 + d_{31}E_3,$$

$$\varepsilon_2 = S_{11}^E \sigma_2 + S_{12}^E \sigma_1 + d_{31}E_3, \quad (4)$$

$$D_3 = \varepsilon_{33}^T E_3 + d_{31}(\sigma_1 + \sigma_2), \quad (5)$$

where ε_1, ε_2 are the relative strains in the meridian and circumferential directions; S_{1i}^E ($i = 1,2$) are the elastic flexibilities of ceramics; σ_i are the mechanical stresses.

From equations (4) let us express σ_1 and σ_2:

$$\sigma_1 = [S_{11}^E(1-\mu^2)]^{-1}[\varepsilon_1 + \mu\varepsilon_2 - (1+\mu)d_{31}E_3],$$
$$\sigma_2 = [S_{11}^E(1-\mu^2)]^{-1}[\varepsilon_2 + \mu\varepsilon_1 - (1+\mu)d_{31}E_3], \qquad (6)$$

where $\mu = -S_{12}^E/S_{11}^E$ is the Poisson's coefficient.

For a piezoceramic shell with uneven thickness polarization in accordance with the accepted distributions of polarization (1) and piezomodule $d_{31}(z)$ (2) (See Fig. 69) we take the strains of the piezoceramic shell in the form

$$\varepsilon_1 = \varepsilon_1^{(0)}\left(1 + \frac{2z}{h}\right) + z\kappa_1 \quad \varepsilon_2 = \varepsilon_2^{(0)}\left(1 + \frac{2z}{h}\right) + z\kappa_2 \qquad (7)$$

where ε_i^0 ($i = 1,2$), κ_i are the relative strains and changes of the main curvatures of the middle shell surface.

Taking into account (2), (3) and (7), we write (6) as

$$\sigma_1 = \frac{1}{S_{11}^E(1-\mu^2)}\left\{\varepsilon_1^{(0)} + \mu\varepsilon_2^{(0)} - \frac{1+\mu}{2}d_{31}^{(0)}E_3^{(0)} + z\left[\frac{2}{h}(\varepsilon_1^{(0)} + \mu\varepsilon_2^{(0)}) + \right.\right.$$
$$\left.\left. + \kappa_1 + \mu\kappa_2 - \frac{1+\mu}{h}d_{31}^{(0)}E_3^{(0)} - \frac{1+\mu}{2}d_{31}^{(0)}E_3^{(1)}\right] - z^2\frac{1+\mu}{h}d_{31}^{(0)}E_3^{(1)}\right\}, \qquad (8)$$

$$\sigma_2 = \frac{1}{S_{11}^E(1-\mu^2)}\left\{\varepsilon_2^{(0)} + \mu\varepsilon_1^{(0)} - \frac{1+\mu}{2}d_{31}^{(0)}E_3^{(0)} + z\left[\frac{2}{h}(\varepsilon_2^{(0)} + \mu\varepsilon_1^{(0)}) + \right.\right.$$
$$\left.\left. + \kappa_2 + \mu\kappa_1 - \frac{1+\mu}{h}d_{31}^{(0)}E_3^{(0)} - \frac{1+\mu}{2}d_{31}^{(0)}E_3^{(1)}\right] - z^2\frac{1+\mu}{h}d_{31}^{(0)}E_3^{(1)}\right\}. \qquad (9)$$

Calculating the internal mechanical forces in the shell T_1, T_2 and bending moments M_1, M_2 through the integrals of σ_1 and σ_2 and considering (8), (9), we obtain the following electroelasticity relations:

$$T_1 = D_T\left[\varepsilon_1^{(0)} + \mu\varepsilon_2^{(0)} - \frac{1+\mu}{2}d_{31}^{(0)}E_3^{(0)} - \frac{h}{12}(1+\mu)d_{31}^{(0)}E_3^{(1)}\right],$$

$$T_2 = D_T\left[\varepsilon_2^{(0)} + \mu\varepsilon_1^{(0)} - \frac{1+\mu}{2}d_{31}^{(0)}E_3^{(0)} - \frac{h}{12}(1+\mu)d_{31}^{(0)}E_3^{(1)}\right], (10)$$

$$M_1 = D_M\left[\frac{2}{h}(\varepsilon_1^{(0)} + \mu\varepsilon_2^{(0)}) + \kappa_1 + \mu\kappa_2 - (1+\mu)d_{31}^{(0)}\left(\frac{E_3^{(0)}}{h} + \frac{E_3^{(1)}}{2}\right)\right],$$

$$M_2 = D_M\left[\frac{2}{h}(\varepsilon_2^{(0)} + \mu\varepsilon_1^{(0)}) + \kappa_2 + \mu\kappa_1 - (1+\mu)d_{31}^{(0)}\left(\frac{E_3^{(0)}}{h} + \frac{E_3^{(1)}}{2}\right)\right], (11)$$

where $D_T = h[S_{11}^E(1-v^2)]$, $D_M = h^3/[12S_{11}^E(1-\mu^2)]$.

The value $E_3^{(1)}$ in the electroelasticity relations (11) is determined from equation (5), which in view of expressions (2), (3) and (7) will take the form

$$D_3^{(0)}\left(\frac{1}{2} + \frac{z}{h}\right) = \varepsilon_{33}^T E_3^{(0)} + \frac{d_{31}^{(0)}}{2S_{11}^E(1-\mu)}(\varepsilon_1^{(0)} + \varepsilon_2^{(0)} - d_{31}^{(0)}E_3^{(0)}) +$$

$$+ z\left\{\varepsilon_{33}^T E_3^{(1)} + \frac{d_{31}^{(0)}}{2S_{11}^E(1-\mu)}\left[\frac{2}{h}(\varepsilon_1^{(0)} + \varepsilon_2^{(0)}) + \kappa_1 + \kappa_2 - \right.\right.$$

$$\left. - \frac{2}{h}d_{31}^{(0)}E_3^{(0)} - d_{31}^{(0)}E_3^{(1)}\right] + \frac{d_{31}^{(0)}}{hS_{11}^E(1-\mu)}(\varepsilon_1^{(0)} + \varepsilon_2^{(0)} - d_{31}^{(0)}E_3^{(0)})\right\} +$$

$$+ z^2\left\{-\frac{(d_{31}^{(0)})^2 E_3^{(1)}}{S_{11}^E h(1-\mu)} + \frac{d_{31}^{(0)}}{S_{11}^E h(1-\mu)}\left[\frac{2}{h}(\varepsilon_1^{(0)} + \varepsilon_2^{(0)}) + \kappa_1 + \kappa_2 - \right.\right.$$

$$\left. - \frac{2}{h}d_{31}^{(0)}E_3^{(0)} - d_{31}^{(0)}E_3^{(1)}\right]\right\} - z^3\frac{(d_{31}^{(0)})^2 E_3^{(1)}}{h^2 S_{11}^E(1-\mu)}. \qquad (12)$$

In the right-hand part of equality (12) the square of the piezomodule $d_{31}^{(0)}$ has the order of 10^{-20} C²/N², and $z \le 10^{-3}$; hence, we proceed to disregard the members with z^2 and z^3.

Equating the coefficients at the same powers of z in the left- and right-hand parts of the equality (12), let us determine D_3^0 and $E_3^{(1)}$:

$$D_3^{(0)} = 2\varepsilon_{33}^T E_3^{(0)} + \frac{d_{31}}{S_{11}^E(1-\mu)}(\varepsilon_1^{(0)} + \varepsilon_2^{(0)} - d_{31}^{(0)}E_3^{(0)}). \qquad (13)$$

Substituting (13) into (12), we find

$$E_3^{(1)} = \{2E_3^{(0)}[2\varepsilon_{33}^T S_{11}^E(1-\mu) + (d_{31}^{(0)})^2] - d_{31}^{(0)}[2(\varepsilon_1^{(0)} +$$

$$+ \varepsilon_2^{(0)}) + h(\kappa_1 + \kappa_2)]\}/\{h[2\varepsilon_{33}^T S_{11}^E(1-\mu) - (d_{31}^{(0)})^2]\}. \qquad (14)$$

For PZT-4 piezoceramics at $\varepsilon_{33}^{T} = 1.32 \cdot 10^{-4}$ F/m, $S_{11}^{E} = 1.5 \cdot 10^{-11}$ m/N, $d_{31}^{0} = -10^{-10}$ C/N we have $2\varepsilon_{33}^{T} S_{11}^{E} (1-\mu) = 2.6 \cdot 10^{-19}$ C/N² and $(d_{31}^{0})^{2} = 0.1 \cdot 10^{-19}$ C²/N²; therefore, (14) can be simplified:

$$E_3^{(1)} = \frac{4E_3^{(0)}\varepsilon_{33}^{T}S_{11}^{E}(1-\mu) - d_{31}^{(0)}[2(\varepsilon_1^{(0)} + \varepsilon_2^{(0)}) + h(\kappa_1 + \kappa_2)]}{2h\varepsilon_{33}^{T}S_{11}^{E}(1-\mu)} .(15)$$

Formula (15) includes E_3^0, ε_i, κ_i (i = 1, 2), which make up (11) as well. Estimating the contribution of each value in (11) and disregarding small members, we obtain simpler expressions for PZT-4 piezoceramic shells:

$$\mathbf{T}_1 = \mathbf{D_T}[\varepsilon_1^{(0)} + \mu\varepsilon_2^{(0)} - \gamma(\kappa_1 + \kappa_2) - E_{3T}^{(0)}],$$

$$T_2 = D_T[\varepsilon_2^{(0)} + \mu\varepsilon_1^{(0)} - \gamma(\kappa_1 + \kappa_2) - E_{3T}^{(0)}],$$

$$M_1 = D_M\left[\frac{2}{h}(\varepsilon_1^{(0)} + \varepsilon_2^{(0)}) + \kappa_1 + \mu\kappa_2 - E_{3M}^{(0)}\right],$$

$$\mathbf{M}_2 = \mathbf{D_M}\left[\frac{2}{h}(\varepsilon_1^{(0)} + \varepsilon_2^{(0)}) + \kappa_2 + \mu\kappa_1 - E_{3M}^{(0)}\right], \qquad (16)$$

where $E_{3T}^{(0)} = \frac{2}{3}E_3^{(0)}(1+\mu)d_{31}^{(0)}$, $E_{3M}^{(0)} = \frac{3}{h}E_3^{(0)}(1+\mu)d_{31}^{(0)}$,

$\gamma = h(1+\mu)(d_{31}^{(0)})^2 /[24S_{11}^{E}\varepsilon_{33}^{T}(1-\mu)]$.

Let us consider a piezoelectric IR modulator in the form of a console-mounted homogeneous plate [Bogomol'nyi V.M. *Measurement Techniques.* 1992. No.9, p.39] with a length $\ell = 35$ mm and cross section 5 × 0.35 mm ($h = 0.35$ mm). The bend of the free end of the console piezoceramic plate is determined by the formula $f = \ell^2\kappa_1/2$, derived from the following conditions: $T_1 = M_1 = 0$ and $\kappa_1 = \text{const}$ along the length of the console plate ($T_2 = M_2 = \varepsilon_2 = \kappa_2 = 0$) [Timoshenko S.P., Lessels A.M., Mech A.S. Applied Elasticity. East Pittsburg: Westinghouse Techn. Night School Press. 1925, p. 91].

The first and third equations (16) give us

$$\varepsilon_1^{(0)} - \gamma\kappa_1 - E_{3T}^{(0)} = 0, \quad 2\varepsilon_1^{(0)}/h + \kappa_1 - E_{3M}^{(0)} = 0 .(17)$$

Let us express $\varepsilon_1^{(0)}$ in the first equation (17) through κ_1, determine κ_1 and the bend of the free end of the console f by the formula

$$f = \frac{5}{6}\frac{\ell^2}{h}E_3^{(0)}(1+\mu)d_{31}^{(0)} \, .$$

At $E_3^{(0)} = 50$ V/mm the bend of a PZT-4 ceramic plate will amount to ff = 2.46 μm.

Thus, the manufacture of piezoelectronic devices (sensors, resonators, surface wave acoustoelectronics) requires the knowledge of how polarization is distributed along their thickness, which cannot be determined by simple engineering techniques. This can be attained by the following inverse method: taking into account the character of polarization changes, we calculate the bend of the sample (a console homogeneous piezoceramic plate); the result set against experiment can demonstrate the real distribution of polarization.

17. Electroelasticity relations for piezoceramic plates and shells

At the inverse piezoeffect relative strains directed towards the thickness of a rectangular parallelepiped-like piezoelement $\varepsilon_3 = \Delta\ell_3 / \ell_3$ are determined by the formula

$$\Delta\ell_3 / \ell_3 = d_{33}E_3 + s_{33}^E F_3 / (\ell_1\ell_2) \, ,$$

where d_{33} is the piezomodule; E_3 is the electric field strength; s_{33}^E is the compliance; F_3 is the external force applied in the direction of the pre-polarization of piezoceramics; ℓ_1, ℓ_2, ℓ_3 are the dimensions of the piezo-element.

The difference of electric potentials U_{el} on the electrodes of the piezo-element under the force F_3 (direct piezoeffect),

$$U_{el} = g_{33}\frac{F_3}{\ell_1\ell_2}\ell_3 \, ,$$

where g_{33} is the electromechanical constant.

In ignition systems of carburetor engines and piezoelectric lighters the mechanical compression stress $1.4 \cdot 10^7$ N/m2 (14 MPa) creates an electric energy of 0.5 J/cm3.

Strains in the plane of a rectangular piezoelement can be determined by the following expression

$$\frac{\Delta\ell_1}{\ell_1} = \frac{\Delta\ell_2}{\ell_2} = d_{31}\frac{U_{el}}{\ell_3} + s_{11}^E\frac{F_3}{\ell_1\ell_2}.$$

The resonance frequency of the rectangular piezoelement at $\ell \geq 2.5\ell_1\ell_2$

$$f_{res} = \frac{1}{2\ell_3\sqrt{\rho s_{33}^E}},$$

where ρ is the density of piezoceramics.

The axis displacement U and the bend of the piezoceramic cylindrical shell (polarized along thickness, free from external mechanical loads) at electrostatic excitation is determined by formulas

$$U = d_{31}E_z^{(0)}\frac{S_0}{2}, \quad W = d_{31}E_z^{(0)}R_0, \quad E_z^{(0)} = -\frac{2V_0}{h},$$

where U, W are the displacements of the middle shell surface in the axial and radial directions; R_0 is the radius of the middle shell surface; $V = \pm V_0$, V_0 are the electric potentials on the shell surfaces with coordinates $z = \pm h/2$; h, s are the thickness and length of the shell (the origin of coordinates is in the center of the shell; the center is rigidly mounted, i.e. with zero axial displacement).

When the butt-ends of the shell are free, the tensing forces in the axial direction are zero, and circumferential stresses σ_2

$$\sigma_2 = \mu d_{31}E_z^{(0)}/s_{11}^E(1-\mu^2),$$

where μ is the Poisson's coefficient.

Dynamic displacements and strains at the first resonance frequencies approximately equal the static ones, multiplied by the quality of piezoceramics Q_M.

A more precise calculation of the dynamic axial displacements of the middle surface of the piezoceramic cylindrical shell can be done by the formulas $U^N = A\sin R\tilde{s}$, $U^N = U/R_0$, $\tilde{s} = s/R_0$,

$$A = \frac{(1-\lambda-\mu)E_z^{(0)}(1+\mu)d_{31}}{(1-\lambda-\mu^2)R\cos(Rs_0/2)},$$

$$R^2 = \frac{\lambda}{[1-\mu^2/(1-\lambda)]}, \quad \lambda = \frac{\rho h R_0^2\omega^2}{D_N}; \quad D_N = \frac{h}{s_{11}^E(1-\mu^2)}.$$

The dynamic displacements $W^N = W/R_0$

$$W^N = \frac{(1+\mu)d_{31}E_z^{(0)}}{1-\lambda}\left[1 - \frac{(1-\lambda-\mu)}{(1-\lambda-\mu^2)}\frac{\mu\cos R\tilde{s}}{\cos(R\tilde{s}_0/2)}\right],$$

$$W^N\bigg|_{\tilde{s}=\tilde{s}_0/2} = \frac{(1-\mu^2)}{(1-\lambda-\mu^2)}d_{31}E_z^{(0)} \text{ at } \tilde{s} = \frac{\tilde{s}_0}{2}.$$

The circumferential dynamic stresses near the free end of the shell can be calculated by the formula

$$\sigma_2\bigg|_{\tilde{s}=\tilde{s}_0/2} = \frac{\lambda d_{31}E_z^{(0)}}{(1-\lambda-\mu^2)s_{11}^E}.$$

In order to increase strength of brittle ceramics, which tenses poorly, to widen the operating frequency range and ensure electromechanical coupling, piezoceramics is reinforced with metal, which creates preliminary compression mechanical stresses. To avoid tension stresses, preliminary compression is achieved by one third to maximum operating load.

Let us consider a cylindrical piezoceramic shell with a temperature-preloaded thin metal shell mounted on it (Fig. 74). There is a preliminary constant contact pressure $\bar{q}_n$ between separate layers, which causes circumferential stresses $\bar{\sigma}_\varphi$ and radial stresses σ_z, determined by the following formulas [Armenakas A.E. Effect of initial strains on the oscillations of free-mounted cylindrical shells. AIAA Journ. 1964. Vol.2. No.9, p.115-122]:

$$\bar{\sigma}_z = \frac{\bar{q}_n(R+h)^2}{(R+h)^2-R^2} + \frac{\bar{q}_n(R+h)^2}{(R+h/2+z)[(R+h)^2-R^2]},$$

$$\bar{\sigma}_\varphi = \frac{\bar{q}(R+h)^2}{(R+h)^2-R^2} - \frac{\bar{q}_n(R+h)^2R^2}{(R+h/2+z)^2[(R+h)^2-R^2]},$$

where R is the radius of the inner surface of the piezoshell; z varies within $-h/2 \le z \le h/2$ (the positive direction of the z axis coincides with the unit vector $\bar{n}$ (see Fig. 74)

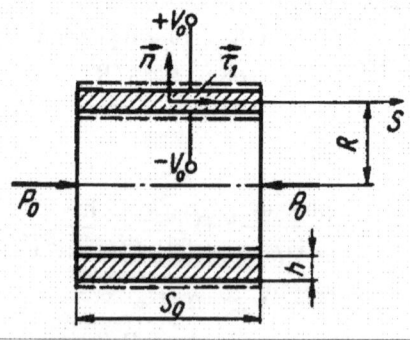

Fig. 74. Cylindrical piezoceramic shell

Since $h/2 \ll R$, expression (9) can be simplified, whereas

$$\sigma_z = -\overline{q}_n \left(\frac{1}{2} + \frac{z}{h} \right).$$

A pre-stressed piezoceramic shell will be characterized by the following electroelasticity relations:

$$N_S = D_N \left[\varepsilon_S^{(0)} + \mu\varepsilon_\varphi^{(0)} - (1+\mu)\left(d_{31}E_z^{(0)} - \frac{\overline{q}_n s_{13}^E}{2} \right) \right],$$

$$N_\varphi = D_N \left[\varepsilon_\varphi^{(0)} + \mu\varepsilon_S^{(0)} - (1+\mu)\left(d_{31}E_z^{(0)} - \frac{\overline{q}_n s_{13}^E}{2} \right) \right],$$

$$M_S = D_M (\kappa_S + \overline{\mu}\kappa_\varphi) + A(\overline{q}_n),$$

$$M_\varphi = D_M (\kappa_\varphi + \overline{\mu}\kappa_S) + A(\overline{q}_n),$$

$$A(\overline{q}_n) = \frac{h^2 \overline{q}_n}{12 s_{11}^E (1-\mu)} \left\{ \frac{k_p^2}{2d_{31}(1-k_p^2)} [2d_{31}s_{11}^E - (1-\mu)s_{11}^E d_{33}] + s_{13}^E \right\}.$$

Metal-reinforced piezoceramic cylindrical shells

To increase durability and electromechanical coupling, piezoceramics is reinforced with metal. It can be achieved due to temperature difference, with preliminary compression stresses created in ceramics.

The dependence of electromechanical energy transformation (determining the acoustic power of a piezoceramic radiator) on static compression stress in various piezoceramics is given in Fig. 75. It shows that preliminary mechanical compression increases the coefficient of electromechanical coupling by 2.6 times.

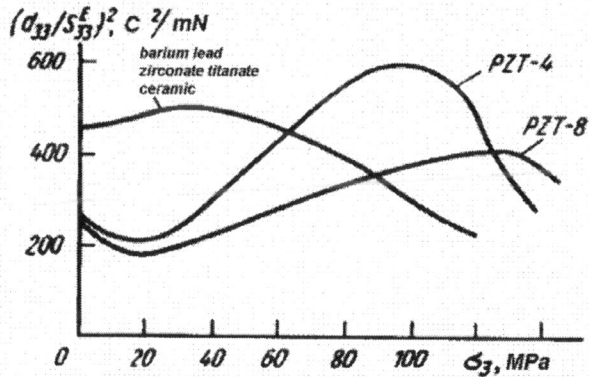

Fig.75. Curves of the changes in the piezomodule depending on compression stresses

Since the electromechanical properties and temperature behaviour of piezoceramics considerably depend on the stress-strain state. We consider piezoceramic shells in view of preliminary mechanical stresses.

Let us consider the oscillations of a cylindrical piezoceramic shell under initial axial forces P_0, which corresponds to the preliminary compression of the shell by a metallic bolt and rigid discs in contact with shell ends. The outer surfaces of the piezoceramic shell $z = \pm h/2$ (h is the thickness of the shell) are supplied with electric potentials $V = V_0 e^{i\omega t}$.

The equations of the movement of the cylindrical shell in view of preliminary compression have the form

$$
\left.
\begin{array}{l}
\dfrac{dT_1^A}{ds} = -U^N \lambda, \quad N_1^N = \dfrac{dM_1^N}{d\widetilde{s}}, \\[2mm]
\dfrac{dN_1^N}{d\widetilde{s}} - T_2^N + \overline{T}_1^N \dfrac{d^2 w^N}{d\widetilde{s}^2} = -\lambda w^N.
\end{array}
\right\}
\tag{1}
$$

Equations (1) use the following notations:

$$
\widetilde{s} = \frac{s}{R_0}, \quad T_1^N = \frac{T_1}{D_N}, \quad T_2 = \frac{T_2}{D_N}, \quad N_1^N = \frac{N_1}{D_N}, \quad \overline{T}_1^N = \frac{P_0}{2\pi R_0 D_N},
$$

$$
M_1^N = \frac{M_1}{R_0 D_N}, \quad U^N = \frac{U}{R_0}, \quad W^N = \frac{w}{R_0}, \tag{2}
$$

$$
\lambda = \frac{R_0^2 h \rho \omega^2}{D_N}, \quad D_N = \frac{h}{s_{11}^E (1-\mu^2)},
$$

where s is the linear coordinate in the axial (meridian) direction; R_0 is the radius of the middle shell surface.

Electroelasticity relations for a piezoceramic shell polarized along thickness at axisymmetrical oscillations have the form

$$
\left.
\begin{array}{l}
T_1^N = \varepsilon_1 + \mu \varepsilon_2 - E_0, \quad T_2 = \varepsilon_2 + \mu \varepsilon_1 - E_0, \\[2mm]
M_1 = D_M \kappa_1, \quad M_2 = \mu M_1, \\[2mm]
E_0 = (1+\mu) d_{31} E_z^{(0)}, \quad E_z^{(0)} = -\dfrac{2V}{h}.
\end{array}
\right\}
\tag{3}
$$

We express the strains through displacements

$$\varepsilon_1 = \frac{dU}{ds}, \quad \varepsilon_2 = \frac{w}{R}, \quad \kappa_1 = -\frac{d^2w}{ds^2}. \qquad (4)$$

Let us express T_2^N from equation (3) and N_1^N from the second equation (1) in view of (3), (4) through displacements and substitute the results into the third equation (1)

$$T_2^N = \omega^2 + \mu \frac{dU^N}{ds} - E_0, \qquad (5)$$

$$\qquad (6)$$

Let us determine $dU^N/d\widetilde{s}$ from the first equation (3) through w and substitute into (6)

$$\frac{dU^N}{d\widetilde{s}} = E_0 + T_1^N - \mu w^N, \qquad (7)$$

$$m \frac{d^4 w^4}{d\widetilde{s}^4} - \overline{T}_1^N \frac{d^2 w^N}{d\widetilde{s}^2} + (1 - \lambda - \mu^2) w^N - (1 - \mu) E_0 = -\mu T_1^N.$$

Expressing T_1^N from the latter equation and substituting it into the first equation, we have

$$T_1^N = m_0 \frac{d^4 w^N}{d\widetilde{s}^4} + a \frac{d^2 w^N}{d\widetilde{s}^2} + b w^N + e E_0, \qquad (8)$$

$$m_0 = -\frac{m}{\mu}, \quad a = \frac{\overline{T}_1^N}{\mu}, \quad b = \frac{\mu^2 - 1 + \lambda}{\mu}, \quad c = \frac{1 - \mu}{\mu},$$

$$m_0 \frac{d^5 w^N}{d\widetilde{s}^5} + a \frac{d^3 w^N}{d\widetilde{s}^3} + b \frac{dw^N}{d\widetilde{s}} = -\lambda U^N.$$

We differentiate the latter expression by S

$$m_0 \frac{d^6 \omega_N}{d\widetilde{s}^6} + a \frac{d^4 \omega^N}{d\widetilde{s}^4} + b \frac{d^2 w^N}{d\widetilde{s}^2} = -\lambda \frac{dU^N}{d\widetilde{s}}. \qquad (9)$$

It follows from (7) that

$$\lambda T_1^N + \mu \lambda w^N - \lambda E_0 = -\lambda \frac{dU^N}{d\widetilde{s}}. \qquad (10)$$

Equating the left-hand parts of equations (9) and (10), considering (8), we obtain the main equation of the problem we set:

$$\frac{d^6 w^N}{d\widetilde{s}^6} + n \frac{d^4 w^N}{d\widetilde{s}^4} + d \frac{d^2 w^N}{ds^2} + e w^N = f_0(\lambda, E_0), \qquad (11)$$

$$n = \frac{a - \lambda m}{m_0}, \quad d = \frac{b - \lambda a}{m_0}, \quad e = \frac{\lambda(\mu - b)}{m_0}, \quad f_0 = \frac{\lambda E(1 + c)}{m_0}.$$

The solution of equation (11) is found by analyzing the roots of its characteristic equation

$$x^6 + nx^4 + ds^2 + e = 0, \qquad (12)$$

which by sequential substitutions $y_* = x^2$, $y_* = y - n/3$ is brought to a cubic equation

$$y^3 + py + q = 0,$$

$$p = d - n^2/3; \quad q = e - \frac{nd}{3} + \frac{n^2}{9} - \frac{n^3}{27}.$$

The real and complex roots of characteristic equation (12) are determined depending on the sign of the D discriminant:

$$D = \left(\frac{p}{3}\right)^3 + \left(\frac{q}{2}\right)^2.$$

Taking into account the parity of the w^N function, one solution of equation (12) can be obtained by A.N.Krylov's functions:

$$w^N = C_1 ch(x_1 \tilde{s}) + C_2 ch\alpha \tilde{s} \cos\beta \tilde{s} + C_3 sh\alpha \tilde{s} \sin\beta \tilde{s} + \frac{f_0}{e},$$

$$\alpha = \frac{b}{2\beta}, \quad \beta = \left(-\frac{\bar{a}}{2} + \frac{\sqrt{\bar{a}^2 + \bar{b}^2}}{2}\right), \quad \bar{a} = -\frac{A + B}{2}; \quad \bar{b} = \frac{A + B}{2}\sqrt{3},$$

$$A = \left(-\frac{q}{2} + \sqrt{D}\right)^{1/3}, \quad B = \left(-\frac{q}{2} - \sqrt{D}\right)^{1/3},$$

where C_1, C_2, C_3 are the integration constants; x_1 is the real solution of the cubic equation.

We consider a cylindrical metal-prestressed piezoceramic shell. There is an initial contact pressure $\bar{q}_n$ between separate layers. The outer surfaces of the thickness-polarized shell with coordinates $z = \pm h/2$ are supplied with electric potentials $V = \pm V_0 e^{i\omega t}$.

At a certain initial difference between the outer diameter of the piezoceramic shell and the inner diameter of the metal shell, the value of the constant contact pressure $\bar{q}_n$ (which occurs due to temperature difference) is determined by the methods of the flat theory of elasticity.

When a piezoceramic element oscillates, there is a supplementary alternating normal pressure $\widetilde{q}_n^{(0)}$ on the surface of the contact, which depends on the value of radial displacements and the rigidity of the metal shell.

The strain of the middle surface of the metal shell in the circumferential direction is determined by the formula

$$\varepsilon_2^M = \sigma_{2M} / E_M = w / R_M, \quad (13)$$

where σ_{2M}, E_M, R_M are the circumferential stresses, modulus of elasticity and radius of the middle surface of the metal shell.

From (13) in view of the relation $\sigma_{2M} = \widetilde{q}_n^{(0)} R_M / h_M$ (h_M is the metal thickness) we determine $\widetilde{q}_n^{(0)}$

$$\widetilde{q}_n^{(0)} = E_M h_M w / R_M. \quad (14)$$

Under external load $\overline{q}_n$, averaged to the outer surface of the cylindrical shell, there is an internal preliminary compression force $\overline{T}_2$, which is determined by the formulas

$$\overline{T}_2(\overline{q}_n, z) = \int_{-h/2}^{h/2} \sigma_2 \, dz, \quad (15)$$

$$\overline{\sigma}_2 = \frac{-\overline{q}_n (R + h/2)^2}{(R + h/2)^2 - (R - h/2)^2} - \frac{\overline{q}_n (R + h/2)^2 (R - h/2)^2}{(R + z)[(R + h/2)^2 - (R - h/2)^2]}.$$

Let us use a thickness-averaged value of $\overline{T}_2$ for calculation:

$$\overline{T}_2 = -\overline{q}_n R \left(1 + \frac{h}{2R}\right). \quad (16)$$

The stress-strain state of a pre-stressed piezoceramic shell at harmonic oscillations consists of the constant and variable components. The total circumferential forces in the piezoceramic shell T_2 are

$$T_2 = \overline{T}_2 + \widetilde{T}_2(\omega), \quad (17)$$

where $\overline{T}_2$ is calculated by formulas (15), (16), and $\widetilde{T}_2(\omega)$ depends on E_3, $\overline{q}_n$, $\widetilde{q}_n^{(0)}$ and is determined in this paragraph.

Firstly, we consider the statistical calculation, followed by the dynamic problem in view of the preliminary mechanical load of the piezoceramic shell. The equations of movement and electroelasticity will have the form

$$\frac{dT_1}{d\widetilde{s}} - \overline{T}_2 \frac{dw^N}{d\widetilde{s}} = -\lambda U^N, \quad (18)$$

$$\frac{dN_1^N}{d\tilde{s}} - T_2^N - \tilde{q}_n(w^N) = -\lambda w^N, \quad (19)$$

$$\frac{dM^N}{d\tilde{s}} - \overline{T}_2^N \frac{h}{2R} \frac{dw^N}{d\tilde{s}} = N_1^N, \qquad (20)$$

$$\tilde{q}_n(w^N) = \frac{E_M h_M}{D_N} w^N,$$

$$\left.\begin{array}{l} T_1^N = \varepsilon_1^{(0)} + \mu\varepsilon_2^{(0)} - E_0, \\[2mm] T_2^N = \varepsilon_2^{(0)} + \mu\varepsilon_1^{(0)} - E_0, \\[2mm] T_1^N = \dfrac{T_1}{D_N}, \ T_2^N = \dfrac{T_2}{D_N}, \ \overline{T}_2^N = \dfrac{\overline{T}_2}{D_N}, \ N_1^N = \dfrac{N_1}{D_N}, \end{array}\right\} \qquad (21)$$

$$M_1^N = \frac{M_1}{R_0 D_N}, \ \lambda = \frac{R_0^2 h\rho\omega^2}{D_N}, ; \ E_0 = (1+\mu)(d_{33}E_z^{(0)} + \tilde{q}_n^{(0)} s_{13}^E / 2),$$

where $\overline{T}_2^N$ is the constant internal force caused by preliminary pressure $\overline{q}_n = \text{const}$; the value $\tilde{q}_n^{(0)}$ is determined by formula (14).

For a piezoceramic element in the form of a PZT-4 cylindrical shell with a thickness $h = 2$ mm, radius of the middle surface $R_0 = 10$ mm and metal shell with $h_M = 0.5$ mm, $R_M = 11.25$ mm, $E_M = 2.1 \cdot 10^5$ MPa at dynamic displacements of $w = 10$ μm the value of $\tilde{q}_n^{(0)} = 0.1$ MPa.

In piezoceramics preliminary compression stresses caused by temperature difference constitute up to half of the operating load. Since the value of permissible dynamic stresses for PZT-4 is $[\sigma] = 30$ MPa, the pressure $\overline{q}_n$ amounts to $\overline{q}_n = 2 - 3$ MPa, i.e. $\overline{q}_n \gg \tilde{q}_n^{(0)}$.

We express T_2^N from the second relation (21), and N_1^N from (20) through displacements and substitute into equation (19):

$$\left.\begin{array}{l} m\dfrac{d^4 w^N}{d\tilde{s}^4} + \overline{T}_{2(1)}^N \dfrac{d^2 w^N}{d\tilde{s}^2} + (\alpha - \lambda)w^N + \mu\dfrac{dU^N}{d\tilde{s}} = E_0, \\[4mm] m = \dfrac{D_M}{R_0^2 D_N}, \ \overline{T}_{2(1)}^N = \dfrac{\overline{T}_2^N h}{2R_0}, \ \alpha = 1 + \dfrac{E_M h_M}{D_N}. \end{array}\right\} \qquad (22)$$

Determining $dU^N / d\tilde{s}$ from the first equation (21) through displacements and substituting into (22), we obtain

$$m\frac{d^4 w^N}{d\tilde{s}^4} - T_{2(1)}^N \frac{d^2 w^N}{d\tilde{s}^2} + (\alpha - \lambda - \mu^2)w^N + \mu T_1^N = (1-\mu)E_0 .(23)$$

We express T_1^N from this equation through the other unknown functions and differentiate this expression by $\tilde{s}$

$$T_1^N = m_0 \frac{d^4 w^N}{d\tilde{s}^4} + \tilde{a}\frac{d^2 w^N}{d\tilde{s}^2} + \tilde{b}w^N + \tilde{c}, \qquad (24)$$

$$m_0 = -\frac{m}{\lambda}, \ \tilde{a} = -\frac{\overline{T}_{2(1)}^N}{\mu}, \ b = \frac{\alpha - \lambda - \mu^2}{-\mu}, \ \tilde{c} = \frac{(1-\mu)E_0}{\mu},$$

$$\frac{dT_1^N}{d\tilde{s}} = m_0 \frac{d^5 w^N}{d\tilde{s}^5} + \tilde{a}\frac{d^3 w^N}{d\tilde{s}^3} + b\frac{dw^N}{d\tilde{s}}. \quad (25)$$

Substituting (25) into equation (18), we differentiate and obtain the following

$$m_0 \frac{d^6 w^N}{d\tilde{s}^6} + \tilde{a}\frac{d^4 w^N}{d\tilde{s}^4} + (\tilde{b} - \overline{T}_2^N)\frac{d^2 w^N}{d\tilde{s}^2} = -\lambda\frac{dU^N}{d\tilde{s}}. \qquad (26)$$

The first equation (21) gives us $dU^N / d\tilde{s}$

$$\frac{dU^N}{d\tilde{s}} = T_1^N - \mu w^N + E_0 . \qquad (27)$$

Using (24), (27) and (26), we obtain the main equation of the problem:

$$\frac{d^6 w^N}{d\tilde{s}^6} + n\frac{d^4 w^N}{d\tilde{s}^4} + d\frac{d^2 w^N}{d\tilde{s}^2} + \tilde{e}w^N = f^*(\lambda, E_0), \qquad (28)$$

$$\tilde{n} = \frac{\tilde{a} + \lambda m}{m_0}, \ \tilde{d} = \frac{b - \overline{T}_2^N - \lambda\tilde{a}}{m_0}, \ \tilde{e} = \frac{\lambda(\tilde{b} - \mu)}{m_0} f^*(\lambda, E_0) = \frac{\lambda(\tilde{c} + E_0)}{-m_0}.$$

Equation (28) can be solved by the method stated earlier in the paragraph under the following edge conditions:

$$\frac{dw^N}{d\tilde{s}}\bigg|_{\tilde{s}=0} = 0, \ M_1^N = T_1^N = 0 \ \text{at} \ \tilde{s} = \frac{\tilde{s}_0}{2},$$

where $\tilde{s} = 0$ is the coordinate of the shell centre; $\tilde{s}_0$ is the shell length (the second edge condition corresponds to the case when shell ends are free from external mechanical loads).

Let us formulate the directions in which the coefficient of electrome-chanical coupling increases, thus improving the functionality of pie-zotransducers:

1. Preliminary compression. In a cylindrical piezoelement supported by an axial bolt, the compressing stress $\sigma_{cs} = 140$ MPa increases the PZT-4

piezomodule by 1.8 times perpendicular to the action of σ_{cs} (see Fig.75). Similar results have been obtained for rectangular parallelepipeds of various shape (with 0.3 to 3 length/thickness ratio and 0.5 to 1 thickness to height ratio) made of electrostrictive ceramics T-4000. Preliminary compression stresses are from 1/3 to 1/2 permissible cyclic stresses (one should also take into account the dependence of quality on mechanical stresses, see Fig. 76, 77).

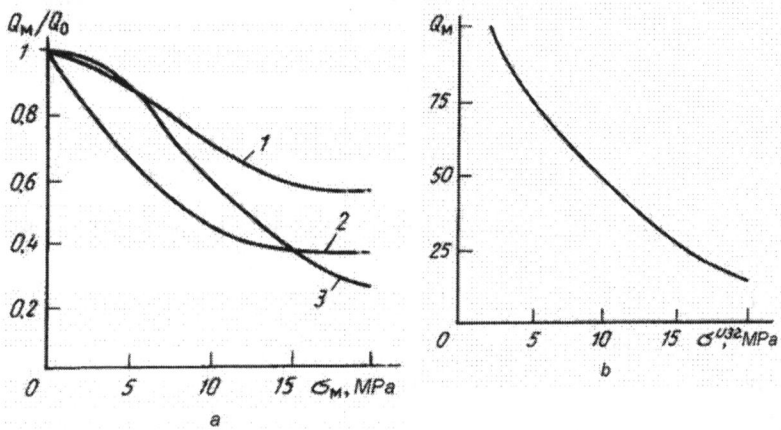

Fig.76. Dependence of quality on mechanical stresses: (a) 1 is the TsTS-23; 2 – TBK-3; 3 – TsTONV-1 (in the Russian notation system); (b) represents the curve of mechanical quality QM for PZT-4

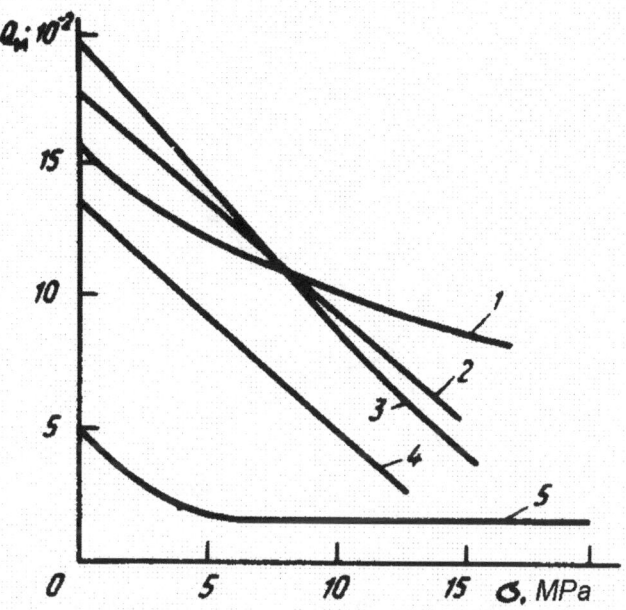

Figure 77. Curves $Q_M \cdot 10^{-2}$ for piezoceramic materials: 1 – PKR-23; 2 – PKR-22; 3 – PKR-8; 4 – PKR-6 (in the Russian notation system); 5 – PZT-4

2. Use of sheer and torsion oscillations. It decreases the deformation energy by several times in comparison with longitudinal oscillations (determined by tension-compression strains).

3. Shape and dimension ratio. For the initial lowest form of oscillations the maximal CEMC is realized when the ratio of the inner diameter of ring piezoceramic plates and their outer diameter is from 0.4 to 0.75 [Adelman N.T., Stavsky Y., Segal E. Axisymmetric vibrations of radially polarized piezoelectric ceramic cylinder. J. Sound And Vibration. 1975. Vol.38. No.2, p.245-254; Contour vibration of clamped piezoelectric annular plates: Analysis for development of an ultrasonic motor. J. Acoust. Soc. Jap. (E). 1990. Vol.11. No.3, p.161-171]. For piezoceramic cylindrical elements polarized along the axis, the CEMC curves depending on the ratio of the outer diameter to the length are given in Fig. 78. As follows from the diagram, CEMC changes by 12-25 times depending on the shape of the piezoelement.

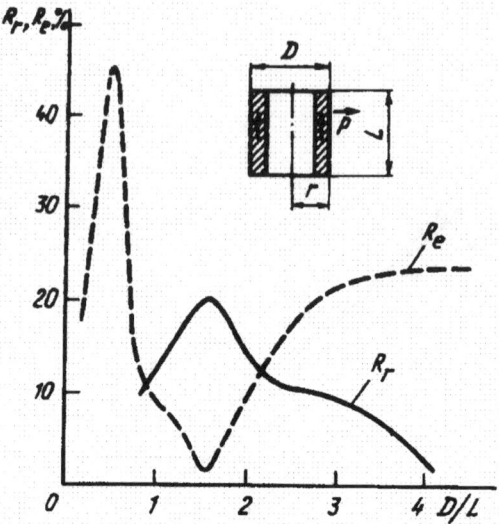

Fig. 78. Coefficients of electromechanical coupling

The source [Lazutkin V.N., Mikhailov A.I. Oscillations of piezoceramic cylinders with finite sizes and height polarization. Sov. Acoust. J. 1976. Vol.22. No.3, p.393-399] contains the results of experiments and calculations of dynamic capacities of piezoceramic cylindrical transducers depending on the ratio of their length and the outer diameter for the first four forms of oscillations (dynamic capacity characterizes electromechanical coupling). By changing the cylinder's dimensions, one can increase the efficiency of energy transformation for each form of eigenfrequencies or eliminate undesirable resonances. CEMC curves, which depend on the ratio of the thickness of a piezoceramic shell and its outer diameter, are given in [Hueter T.F., Bolt R.H. Sonics Techniques for the Use of Sound in Engineering Sciences. New York, Wiley. 1962. Ch.4]. Optimal sizes of piezoceramic cylindrical shells for jet printers are given in [Bogy D.B., Talke F.E. Experimental and theoretical study of wave propagation phenomena in drop-on demand ink-jet devices IBM. J. Res. Develop. Division. 1984. Vol.28. No.3, p.314-321; Sato J., Kawabuchi M., Fukuto A. Dependence of the electromechanical coupling coefficient on the width to thickness ratio of plank-shaped piezoelectric transducers used for electronically scanned ultrasound diagnostic systems. J. Acoust. Soc. Amer. 1979. Vol.6. No.6, p.1609-1611].

For a precise calculation of mechanical stresses and amplitude-frequency characteristics of piezoceramic transducers it is necessary to use the differential equation of oscillations with $\eta\dfrac{dx}{dt}$ (where x is the displacement function). The viscosity coefficient η is calculated by the formula

$$\eta = \frac{E_Y}{\omega Q_M},$$

where E_Y is the Young's modulus, ω is the frequency, $Q_M = 1/tg\delta_{mech}$ is the mechanical quality.

To obtain the maximal amplitude of actuator oscillations, one chooses piezoceramics of rather low quality ($Q \cong 60 - 80$). The application of piezoceramic materials is limited by the increase of heating temperature at harmonic electric excitation (due to dielectric and piezoelectric losses).

As follows from Fig. 79, at the temperature $T \cong 60 \div 65\,°C$ the quality for all variants becomes approximately the same. There is an evident change in the lowest resonance eigenfrequency.

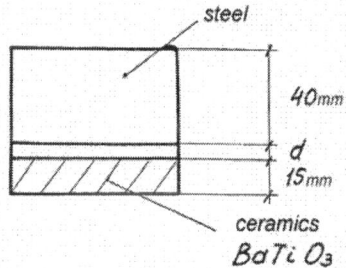

Fig.79. Layout of composite structure (d is the thickness of glue)

The effect of thickness, rigidity and density, as well as the Poisson's coefficient of soft glue substrates was studied in [Kagawa Y., Yamabuchi T. Finite element simulation off composite elastostrictive resonators. "Ultrasonic International Conference Proceedings", 1977, p.138-144]. The experiments and calculations were made for the following types of connective linings (see Table 11).

№№	Connective layer		d thickness, mm	ρ, kg/m3	ν, Poisson's coefficient
I	araldite hardness	121S 930	0.19	1.56	0.4
II	araldite hardness	101S 951	0.02	1.11	0.45

| III | araldite
hardness
filler
quartz sand | 123B
953B | 0.32 | 1.74 | 0.38 |

Table 11

The curves of Q (qualities) of composition elements (see Fig. 79) depending on temperature are shown in Fig. 80.

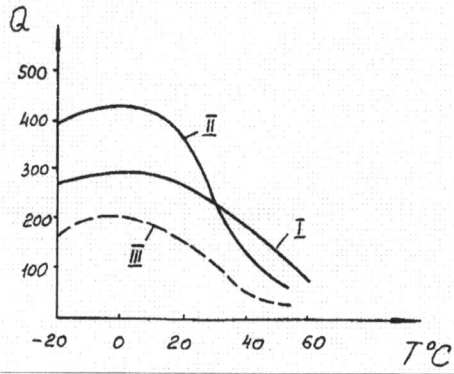

Fig. 80.

As follows from these figures (see Table 11), the quality factor for layer II is approximately 2÷3.5 times larger than for I and III.

The effect of USW oscillations in the baking of piezoceramics improves its physicomechanical characteristics.

The destruction of piezoceramic multi-layer transducers usually occurs along glue connections. This limits the power of sonar radiation (in spite of the preliminary compression of the piezoelement by axial bolts $\sigma_{str} = 15$ MPa) and changes the lowest eigenfrequency by 0.3%). In order to increase cyclic strength, separate piezoelements are assembled not only with glue (rubber-epoxy compositions like araldite), but also by soldering or diffusion welding [Prokopyev S.V. Studying and developing the technology of diffusion welding of ferroelectric soft piezoceramics with metals. Ph. dissertation thesis. Krasnoyarsk: Siberia State Aerospace Institute. 2003].

A disadvantage of bimorph piezoelectric transducers is the destruction of glue connective layers. Using homogeneous flextension piezoceramic plates with uneven thickness polarization, as well as monolithic plates with internal electrodes, llows to increase their dynamic strength [Ogawa T. et al. Monolithic bimorph with internal electrodes. 6th Symp. Ultrasonic

Electronics. Proc. Tokyo 1985. Japn. J. Appl. Phys. 1986. Vol. 25. Suppl. 25-1, p. 25-27].

The shape of a piezoelectric transducer (which increases electromechanical coupling) is determined in [Sato J. et al. Dependence of the electromechanical coupling coefficient on the width to thickness ration of plank-shaped piezoelectric transducers used for electronically scanned ultrasound diagnostic systems. JASA. 1979. Vol.6. No.6, p.1609-1611]. For baking piezoceramic materials and for metal contacts one can employ plasma technologies [Cho J.Y., Yoon K.H., Kim E.S. Influence of thermal stress on quality factor of the layered $Mg0.93Ca0.07TiO3$ – $(Ca0.3Li0.14Sm0.42)TiO3$ system prepared by spark plasma sintering. Japn.J.Appl. Phys. 2003. Vol. 42. No.12, p.7397-7400].

This monograph explains the choice of technological parameters of the electroadhesive combination of various materials (§ 3.1, §3.8). At the initial reversible stage of electrothermal breakdown (at certain threshold values of electric field strength, geometry of roughness and temperature) structural defects are healed and various materials can be connected.

Special technology uses the most reliable and simple thermacoustic refrigerator (freonless) with sensorless control [Shearer T.L. et al. Sensorless control of a thermacoustic refrigerator. JASA. 2004. Vol. 116. No.1, p.288-293; Li B.L. et al. Adaptive thinning of an electrodynamically driven thermacoustic cooler. JASA. 2002. Vol.111, p.1251-1258; Garret S.L. et al. Thermoacoustic refrigerator for space applications. J. Thermophys. Heat Transfer. 1993. Vol. 7, p.595-599; Johnson R.A. et al. Thermacoustic cooling for surface combatants. Naval Eng. J. 2000. Vol. 112, p.335-345].

18. Nonlinear pyroelectric effect. Mechanical and dielectric losses

The Barkhausen heat effect, which is expressed in polarization jumps under an internal spontaneous electric field when the temperature of ferroelectric monocrystals changes, is used for structural analysis [Bogomolov A.A., Malyshkina O.V., Dabizha T.A. Nonlinear pyroeffect in unipolar DTGS crystals. Ferroelectrics. 1996. Vol. 186, p.1].

The time-variable heat flow acting on pyroelectric crystal causes a voltage on its electrodes. The primary and secondary pyroeffects are characteristic under thickness-constant temperature, the tertiary effect is evident when temperature is distributed unevenly. All these physical effects should be taken into account when designing night vision devices, temperature remote sensors and photoreceivers.

Centre-symmetrical (non-polar) DE have also been found to possess the thermopolarization effect. The appearance of electric fields when temperature is distributed unevenly is universal for solids (in metals and semiconductors uneven heating causes thermal EMF).

This paragraph offers a calculation of the electric field and mechanical stresses which occur in a ferroelectric heated at a constant speed. The spontaneous electric internal field is determined by the ratio of the pyroelectric coefficient and dielectric permeability (γ / ε).

The fundamental physical properties of crystals considerably change when their thickness is decreased (optical absorption, dielectric permeability, coercive field, destruction energy). The physics, chemistry and structure of a very thin (0.7-40 μm) surface layer differ from the space layer by two or more orders.

BaTiO3 crystals give a growth surface layer with superior optical properties (regarding the volume). For 75-178 μm BaTiO3 crystals the thickness of the surface layer is 6±2 μm.

At the "surface layer – volume" interface there is an accumulation of the space charge. The domain structure on the surface and in the volume are different. According to various experimental data, the thickness of the BaTiO3 surface layer may change from 0.7 to 25 μm.

It was found that triglycinesulphate (TGS) crystals have up to 45 μm polydomain surface layers [Miller S.L., Nasby R.D. et al. Device modeling of ferroelectric capacitors. J. Appl. Phys. 1990. V. 68. N 12. P. 6463-6477; Pereverzeva L.P. et al. Thermopiezoelectricity in non-centre-symmetrical crystals. Sov. Phys. Solid State. 1992. Vol.34. No.1, p.281-287; Solunsky V.I. Near-surface layers of ion crystals: balance, stationarity, quasistationarity. Sov. Phys. Solid State. 1983. Vol. 25. No.9, p.2696-2701; Bhalla A., Newnham R. Primary and secondary pyroelectricity. Phys. Status Solidi (A). Vol. 58(a), p.19-24].

Mechanical and dielectric losses in ferroelectric crystals in the non-equilibrium state

It has been established experimentally that dielectric or mechanical losses grow linearly with the increase in the speed of the phase transition of the first kind. The measurements of the internal frictions at low frequencies and continuous temperature change have shown that the peak height Q^{-1} changes in direct proportion to the speed of temperature change. This is expressed in the following equations for the internal frictions and the tangent of the dielectric loss angle

$$Q_m^{-1} = 2\pi G \frac{\beta \Delta x_s^2 m}{kT_k \omega},$$

$$tg\delta_m = \frac{4\pi \beta m \Delta P_s^2}{kT_k \omega \varepsilon} \cdot \frac{dn}{dt},$$

where G is the shear module, β is the volume of the critical embryo, Δx_s and ΔP_s are the jumps of spontaneous strain and polarization at the point of phase transition, m is the speed of phase transformation, ω is the frequency, n is the concentration of embryos. It has been shown that these formulas can be applied for various mono- and polycrystal ferroelectrics which undergo phase transitions of the first kind. Isothermal exposure of samples at $T = T_c$ considerably decreases the anomalous internal friction (T_c is the Curie point) [Prasolov B.N., Safonova I.A. Effect of velocity and direction of the phase transition of the second kind on dielectric losses in TGS crystals. Bull. of the Russ. Acad. of Sci. Phys. 1993. Vol.57. No.3, p.47-49].

DE permittivity, dispersion, effective dielectric losses in PZ composites are experimentally and theoretically studied in [Emets Y.P. Modelling of electrophysical characteristics of the dielectric medium with a periodical structure. Sov.Phys. – Techn. Phys. 2004. Vol.74. No.12, p.1-9. http://www.ioffe.rssi.ru/journals/jtf].

Thermal electric fields In ferroelectric crystals under continuous temperature change

Changing the temperature of a ferroelectric crystal brings about elementary acts of repolarization, which are expressed as multiple electric impulses, or Barkhausen jumps.

This paragraph contains the calculation of thermal electric fields in ferroelectric crystals. These fields are caused by temperature gradients under continuous temperature change.

The Barkhausen heat effect is studied at the temperature that depends on time linearly:

$$\Theta = T_0 + b \cdot \tau, \quad (1)$$

where Θ is the temperature of the medium, T_0 is the initial temperature of the medium and the sample, b is the speed at which the medium temperature changes, τ is the time. The calculation was experimentally tested on various $0\times10\times1\div2$ mm TGS-like crystal specimens. This allows us to calculate heat fields using the one-dimensional approximation of a plane-unlimited thin plate. Let us choose a rectangular system of coordinates x_1, x_2, x_3 with the origin in the central plane of the plate symmetry and the positive direction of the x_2 axis coinciding with the vector of spontaneous polarization.

The temperature gradient is considered stationary and in our problem is expressed by

$$\frac{\partial T}{\partial x_2} = \frac{b}{a_{22}} x_2, \quad (2)$$

where a_{22} is the coefficient of heat diffusion.

The temperature gradient causes uneven polarization, which in its turn creates electric fields. We can calculate these fields by the equation of electrostatics

$$\operatorname{div} D = \rho, \quad (3)$$

where D is the electric induction, ρ is the space density of the free charge; and the induction relation in the form

$$D_i = \varepsilon_0 \varepsilon_{ik} E_k + P_{si} + d_{i\chi}\sigma_\chi \ (i, k = 1, 2, 3 \,; \ \chi = 1, 2, ..., 6), \quad (4)$$

where D_i, E_k, P_{si} are the projections of induction, electric field strength and spontaneous polarization vectors on the coordinate axis, ε_{ik} are the components of dielectric permeability, $d_{i\chi}$ are the piezomodules, σ_χ are the components of the mechanical stress tensor, ε_0 is the dielectric constant.

Ignoring electroconductivity and considering that ε_{ik}, $d_{i\chi}$ and the pyroelectric coefficient γ are constant, we transform expressions (3) in view of (4) into

$$\varepsilon_0\varepsilon_{22}\frac{\partial E_2}{\partial x_2} + \gamma\frac{\partial T}{\partial x_2} + d_{2\chi}\frac{\partial \sigma_\chi}{\partial x_2} = 0. \quad (5)$$

For temperature mechanical stresses in the absence of external loads we have

$$\sigma_\chi = \frac{1}{6}\frac{g_\chi bh^2}{a_{22}}\left[1 - 3\left(\frac{x_2}{h}\right)^2\right], \ \chi = 1, 3, 5 \quad (6)$$

where $g_\chi = k_{\chi\lambda}\alpha_\lambda$ (1, 2, 3), $k_{\chi\lambda}$ are the components of the tensor opposite to the tensor of compliance, α_λ are the coefficients of heat expansion, h is the half-thickness of the pyroelectric plate.

The solution of equation (5) in view of (2) and (6) and the edge condition for the electric potential $\varphi(h) = \varphi(-h)$ (the electrodes of the sample are short-circuited) has the form:

$$E_2 = \frac{1}{6} \frac{bh^2}{\varepsilon_0 \varepsilon_{22} a_{22}} \left(\gamma - d_{2\chi} g_\chi\right) \left[1 - 3\left(\frac{x_2}{h}\right)^2\right]. \quad (7)$$

Expression (7) can be represented as a sum of two items, E_2' and E_2'', determined correspondingly

$$E_2' = \frac{1}{6} \frac{bh^2 \gamma}{\varepsilon_0 \varepsilon_{22} a_{22}} \left[1 - 3\left(\frac{x_2}{h}\right)^2\right], \quad (8)$$

$$E_2'' = \frac{1}{6} d_{2\chi} g_\chi \frac{bh^2}{\varepsilon_0 \varepsilon_{22} a_{22}} \left[1 - 3\left(\frac{x_2}{h}\right)^2\right]. \quad (9)$$

The component E_2' expresses the internal electric field strength caused by uneven spontaneous polarization; the component E_2'' describes the contribution of the piezoeffect.

As seen from (8) and (9), the relation of the field strengths is determined by the formula:

$$\frac{E_2''}{E_2'} = -\frac{d_{2\chi} g_\chi}{\gamma}. \quad (10)$$

This relation was calculated for TGS crystals using the known g_χ, elastic flexibilities, piezomodules and heat expansion coefficients. The relation amounted to 18%. Expression (10) is the relation of the secondary and primary pyrocoefficients. Experimentally it was found to be 20%, which is in good accordance with the calculation. As a first approximation, in the TGS case we can ignore E_2'' in comparison with E_2' and analyse experimental data by formula (8). Figure 81(a) features the distribution of the vectors of the thermal electric field strengths, and Fig. 81(b) shows the field as a function of the x_2 coordinate.

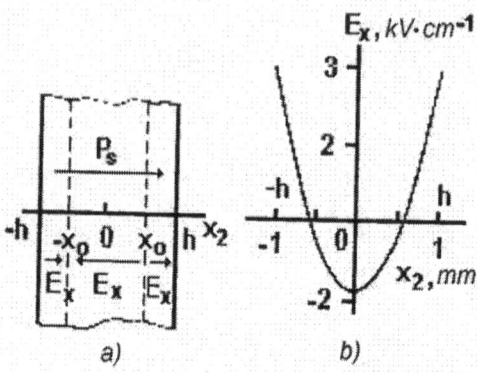

Fig. 81. Distribution of the electric field in a TGS crystal (a is the distribution of the vectors of the thermal electric field strengths, b is the distribution of the electric field along the thickness of the TGS crystal)

The calculation was made for a 2 mm TGS sample, the temperature of the environment $T = 30\,°C$, $\varepsilon_{22} = 60$, $\gamma = -5 \cdot 10^{-4}$ C·m-2K-1, $a_{22} = 2.8 \cdot 10^{-7}$ m2s-1, $b = 0.3$ K·s-1, $h = 1$ mm. As follows from Fig.81(b), the field in the central part of the plate is directed opposite to the polarization vector and in the maximum at $x = 0$ reaches 1.7 kV·cm-1. The field strength considerably exceeds the coercitive field. At points with coordinates $x_0 = \pm 0.58$ mm the field strength becomes zero. Near the plate surface the field strength vector is directed like the polarization vector, and its strength reaches the maximum, 3.4 kV·cm-1 at $x = \pm h$. Thus, the quasistatic heating of a ferroelectric crystal with short-circuited electrodes ($E = 0$) creates in its internal area ($-x_0 < x < x_0$) a rather strong depolarizing field. The field strength E_2', as follows from formula (8), linearly depends both on the heating speed and on the γ / ε relation. There is a quadratic dependence of E_2' on the thickness of the crystal sample.

Barkhausen jumps in ferroelectrics are caused both by the formation of embryos and the jump-like switching of domain boundaries. The jumps that occur when the temperature of ferroelectrics changes can be equated with the formation of domain embryos.

Barkhausen heat effect and thermoelastic mechanical stresses

The Barkhausen heat effect occurs along with the birth of new domains, the polarization in which is antiparallel to the polarization of the monodomain sample. To switch the domain orientation, it is necessary to obtain a depolarizing electric field in the sample at temperature change. Such a field occurs in a monodomain ferroelectric plate of a polar cut with short-circuited electrodes and heated in a gaseous environment with a uniformly changing temperature. This E_2' field is caused by the constancy of spontaneous polarization along thickness. Let us consider the behaviour of E_2'' (which describes the contribution connected with the piezoeffect). When heated at a linearly rising temperature, a crystal plate develops uneven mechanical stresses, described by (6).

A monodomain ferroelectric plate has an uneven piezoelectric polarization

$P_i = d_{i\chi}\sigma_\chi$, $i = 1,2,3$, which creates a bound charge with a density

$$\rho = -\frac{\partial P_i}{\partial x_i}.$$

This charge creates an internal electric field, the strength of which is the solution of the Poisson's equation for the φ potential,

$$\frac{\partial^2 \varphi}{\partial x_i^2} = -\frac{1}{\varepsilon_0 \varepsilon_i} \frac{\partial P_i}{\partial x_i}.$$

Taking into account the edge condition $\varphi(h) = \varphi(-h)$ gives an expression for the internal electric field strength:

$$E_2'' = \frac{1}{6} \frac{d_{2\chi}g_\chi bh^2}{\varepsilon_0 \varepsilon_2 a_{22}} \left[1 - 3\left(\frac{x_2}{h}\right)^2 \right].$$

Disregarding the anisotropy of elastic properties and temperature expansion of the crystal, we have:

$$g_1 = g_3 = \frac{\alpha}{s_{33} + s_{31}}, \quad g_5 = 0. \ (11)$$

For TGS, when the direction of spontaneous polarization coincides with the positive direction of the x_2 axis, in view of (11), we obtain

$$E_2'' = \frac{1}{6} \frac{(d_{21} + d_{23})\alpha bh^2}{(s_{33} + s_{13})\varepsilon_0 \varepsilon_2 a_2} \left[1 - 3\left(\frac{x_2}{h}\right)^2 \right]. \quad (12)$$

A comparison of (8) and (12) shows that E_2' and E_2'' have the same dependence on x_2. Taking into account that $(d_{21} + d_{23}) \approx (T_c - T)^{-1/2}$, $\varepsilon = C/[2(T_c - T)]$, $a = \alpha/(s_{33} + s_{13})$ does not depend on temperature, as well as ignoring the temperature dependence of a_2, we obtain $E_2'' = B'(T_k - T)^{1/2}$, where B' is the constant.

This expression coincides with a similar dependence of E_2' on temperature.

In each point inside the plate the vector E_2'' opposes E_2'. In points $x_2 = 0$ of the plate with a thickness $2h = 2$ mm, heated at a speed $b = 0.3$ K·s-1 at 30°C the fields correspondingly are $E_2' = -1.7$ kV·cm-1, and $E_2'' = 0.041$ kV·cm-1. The calculation was based on the following physical values: $\varepsilon_0 = 8.85 \cdot 10^{-12}$ F·m-1, $\varepsilon = 60$, $\gamma = -5 \cdot 10^{-4}$ C·m-2K-1, $a_2 = 2.8 \cdot 10^{-7}$ m2·s-1, $d_{21} = -14.7 \cdot 10^{-12}$ C·N-1, $d_{23} = 45.6 \cdot 10^{-12}$ C·N-1, $s_{53} = 9.75 \cdot 10^{-11}$ m2·N-1, $s_{13} = -2.9 \cdot 10^{-11}$ m2·N-1, $\alpha = 27 \cdot 10^{-6}$ K-1.

In TGS crystals internal electric fields caused by mechanical stresses can be disregarded in this case. In other ferroelectrics the influence of mechanical stresses on the state of the domain structure and the field strength E_2'' may become determinant (for pyroelectrics with d_{21} and d_{23} an order larger than in TGS).

19. Spin and polarization echo processors

If digital techniques and devices fully meet the requirement of fast response, the increase in their speed is connected with decreasing minimal topological sizes, apparently down to submicron technology. The planarity of the contemporary LSI technology considerably limits its connectivity, with connections occupying up to 50% of the crystal area. These, as well as other, causes place a limit of existing technology by integration and speed, drawing LSI designers' attention to other physical principles (e.g. the Josephson's effect).

Solid-state electronics is an alternative to digital technology. It is based on homogeneous physical environments and non-circuit, rather than structured, principles. In this case the role of circuitry is played by physical processes characterized by the use of dynamic non-homogeneities for storing and processing information.

Analog devices have advantages compared to digital devices, because they offer simultaneous access to many "memory cells", as well as parallel operations like multiplication, but with a simpler physics.

Typical examples of functional electronics are memory devices built on cylindrical magnetic domains, charge coupled devices, surface acoustic wave (SAW) devices, optical coherent and non-coherent processors, as well as combined acoustooptic and acoustoelectronic devices, etc.

Research into transition processes at resonance interactions of various physical fields with materials lead to the discovery of the spin echo phenomenon [Hahn E.L. Spin echoes. Phys. Rev. 1950. Vol. 80. No.4, p.580-594; US Pat. 2887673. Impulse nuclear induction technique of spin echo. Hahn E.L. Appl. 13.11.51; Publ. 19.05.59], which gave rise to echo processors [Fernbach S., Proctor W.C. Spin echo memory device. J. Appl. Phys. 1955. Vol. 26, p.170; Anderson A.G., Garwin R.L., Hahn E.L. et al. Spin Echo Serial Storage Memory. J. Appl. Phys. 1955. Vol. 26, p.1324-1338]. Integration in spin echo processors is almost maximal, as determined by the interatomic distance. It is three orders less than the minimal topological size of contemporary integrated circuits and two orders than expected.

The manufacture technology of echo processors is simple enough. Dynamic non-homogeneities used for storing and processing information do not move in space; requirements to space homogeneity and purity of raw materials are small. Precise alignment of geometrical elements is no longer necessary. This gives echo processors a competitive edge in comparison with other devices, which require very pure raw materials and microtechnologies, makes technological processes more reproducible and the yield of usable products.

Echo can be easily described in frequency space, if one uses the known dualism of Fourier transformation. Nonlinear frequency space should contain responses with new temporal positions to a multi-impulse influence, similarly to the appearance of new frequencies in a nonlinear circuit, if there are several EMF with different frequencies. Just like the combination frequencies in the nonlinear circuit are rigidly connected with the acting frequencies, in the nonlinear frequency space the temporal positions of responses are rigidly bound to the temporal position of the acting impulses. A real frequency space can be obtained in the continuum of high-quality oscillators, with frequency characteristics covering a certain range. The role of the oscillator can be played by microparticles (electrons or atom nuclei of condensed media), collective medium excitations (photons, phonons), molecules or even macroscopic particles of the material. The necessary condition is the resonance interaction of the oscillator with the physical (electromagnetic, acoustic, etc.) and rather long eigenfrequencies of the oscillator. These eigenfrequencies should create nonlinear interactions between spectral components of the acting signals that excite the oscillator. A response of such a nonlinear resonance medium, or echo signal, is caused by the co-phase addition of the oscillations in oscillators, and is sometimes called "phased echo".

Since the operation of multiplying signal spectrums (signal convolution, widely used in processing) is dual to the multiplication of signals in the nonlinear circuit, it explains the interest to devices of the kind, capable of short-term memory, convolution or Fourier transformation at radiofrequencies and in real time.

Table 12 gives a notion of capabilities of some echo processors.

Echo type, material	f_0, MHz	$2F$, MHz	T_1, MHz	T_2, MHz	D_0, dB
Nuclear spin					
Co59 TMP (in	215	10	150	25	100
Russian notation)	150-280	5-40	100-200	25	90
Co	45	0.4	104	5×109	100
TMP Ni-Fe-Co					
Fe57 TMP Fe					
Spinel ferrites					
NiFe2O4	70	2.5	3×103	500	60
Li0,5Fe2O4	70	3.5	3×103	700	60
Electron spin					
TiO2 (4.2 K)					
	104	150	–	800	>100
Polarization	10-100	–	1012	100	>100
Ferromagnetic	103-104	100	–	10	>100
Photon					
LaF3+Pr3+ (4.2 K)	6×1014	104	3×108	10-3	>100

Table 12 (Note: f_0 is the resonance frequency; $2F$ is the pass band; T_1 is the time of longitudinal relaxation; T_2 is the time of cross-section relaxation; D_0 is the ratio of the maximally possible response to the eigen noises of the resonance system and the acting material of the echo processor)

The properties of most analog devices can be expressed through the product of the frequency band F and the processing (storage) time T. As to digital devices, it is problematic to estimate their universality and precision, though unattainable for analog equipment. However, these advantages are exhibited only at rather low processing speeds. Let us compare analog and digital processors for convolution or one-dimensional spectral analysis with the frequency band F in the time interval T. A realistic estimate for the required number of samples will be $N = 3FT$, and the product in accordance with the fast Fourier transformation will amount to $2N \log_2 N$ within the time T. If it is possible to make parallel calculations with M multiplies, which multiply Rtimes per second, the operation of this digital system can be expressed by the equation

$$MR = \frac{6FT}{T} \log_2 3FT.$$

Then the main characteristics of analog and digital one-dimensional Fourier transformers can be shown on one frequency-time plane (Fig. 83).

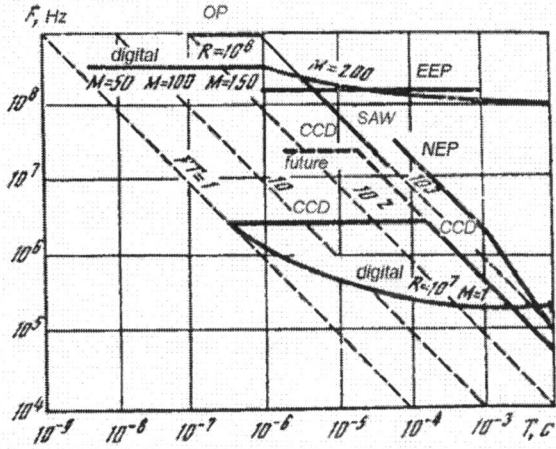

Fig. 82. Compared characteristics of analog and digital processors

SAW and CCD (charge coupled devices) analysers, which operate at linear-frequency modulation (LFM) and LFM-z-transformations correspondingly, reach $FT = 10^3$; the maximal frequency band of the processed signals nears 1 GHz for SAW devices, and 10 MHz for CCD devices. At low frequencies a digital analyzer is superior to any other device ($R = 10^7$ mult/s, $M = 1$).

The upper curve demonstrates the characteristics of future digital analysers, which use ADC (analog-to-digital converters) with 109 samples/s and Mof parallel operating multipliers at $R = 10^8$ mult/s. Increasing the number of ADC and multipliers, it is possible to meet the properties of optical processors; however, one should take into account energy consumption and cost of these devices. For example, to eliminate the difference in the properties of a very elastic optical correlator with a space light modulator (the most powerful to date) and a digital system at $T = 1$ microsecond, one would require 106 multipliers with $R = 10^8$ mult/s. The contemporary properties of echo processors are represented by two curves, for nuclear (NEP) and electron (EEP) echo processors. The characteristics in EEP require helium temperatures; the NEP characteristics are given for room temperature.

The diagram shows that spin NEP and EEP are by no means inferior; one should also take into account the fact that SAW programmable filters based on laminar structures (semiconductor-piezoelectric or Shottky diode-piezoelectric matrix) have a dynamic range 10-20 dB smaller than that given in the diagram. (The dynamic range there is 10FT.) Finally, temperature instability and undesirability of vibrations in SAW devices call for certain measures (thermostats, amortisators), which increases the volume and cost of the device. Thus, echo processors, which are stable towards external mechanical, temperature and radiation influences, have advantages compared to SAW technology.

Traditionally, echo processors have had two drawbacks which limited their application, low dynamic range and bi-polarity. The former has largely been overcome with the appearance of magnetic thin film materials for spin nuclear EP and polarization EP on piezoelectric powders.

One should note the robust character of EP operation in the presence of noise with a priori unknown frequency, intensity and distribution density. It should be also added that EP are only slightly inferior to optimal nonlinear filters and is almost as efficient in processing as an adaptive receiver.

Frequency filters are important elements of mobile wireless phone microprocessors.

LIST OF ABBREVIATIONS

ACS – automatic control system
AE – acoustoelectronics; active element
AEC – autoemission cathode
AEE – autoelectron emission
AEG – acoustoelectronic generator
AFT – automatic frequency tuning
AG – autogenerator
AOD – acoustic-optical device
AOF – acoustic-optical filter
AT – avalanche transistor
BE – Barkhausen effect
BJ – Barkhausen jump
CAEC – carbon autoemission cathode
CCD – charge-coupled devices
CEMC – coefficient of electromechanical coupling
CL – cathode luminescence
CLS – chaotically located sphere
CVC – current-voltage characteristic
DAC – digital-to-analog converter
DC – direct current
DD – damping device
DDR – dielectric disc resonator
DE – dielectrics
DLC – diamond-like carbon
DWT – direct write technology
EEP – electron echo processor
EMF – electromotive force
EMR – electromagnetic resonator; electromechanical resonator; electro-
magnetic radiation
EOC – electron-optical converter
EP – echo processor
FCP – field-controlled piezotransducer
FE – ferroelectrics
FET – field-effect transistor

FL – fluorescent lamp
FTC – frequency temperature coefficient
GGG – gadolinium gallium garnet
HF – high-frequency
HTF – hyperthin film
HTSC – high-temperature superconductivity
IMS – information measuring system
INC – incandescent lamp
IPOS – isolated by porous oxidized silicon
IR – infrared
JJ – Josephson junction
LC – liquid crystal
LED – light-emitting diode
LFM – linear-frequency modulation
LIC – linear integrated circuit
MC – molded channels
MCC – monolithic ceramic circuits
MDM – metal-dielectric-metal
MDS – metal-dielectric-semiconductor
MIS – metal-insulator-semiconductor
MNOS – metal-silicon nitride-silicon oxide-semiconductor structure
MW – microwave
NDR – negative differential resistance
NDT – non-destructive testing
NEA – negative electron affinity
NEP – nuclear echo processor
NR – negative resistance
NTC – negative temperature coefficient of resistance
OFCL – optical fiber communication line
PBST – permeable base submicron transistor
PD – phase detector
PF – positive feedback
PFA – poreforming agent
PM – piezomotor
PS – porous silicon
PSS – phase synchronization system
PT – piezoceramic transformer
PTC – positive temperature coefficient of resistance
PVDF – polyvinylidene fluoride
PZ – piezo
PZT – lead titanate zirconate
RC – remote control

RD – recording device
RSS – relative spectral sensitivity
SAI – semiconductor analogs of inductivity
SAW – surface acoustic waves
SEM – scanning electron microscopy
SIMC – semiconductor integral microcircuit
SN – silicon nitride
SOI – silicon-on-insulator
SQUID – superconducting quantum interference device
STM – scanning tunnel microscope
SWF – shock wave front
TANDEL – temperature autostabilised nonlinear dielectric element
TGS – triglycinesulphate
TM – trasmission mode
USW – ultrasound waves
UV – ultraviolet
VHF – very high frequency
VLSI – very-large-scale integration circuit

REFERENCES

1. Burfoot J.C., Taylor G.W. Polar Dielectrics and their Applications. New York: McMillan Press LTD. 1979.
2. Cady W.G. Piezoelectricity: an Introduction to the Theory and Applications of Electromechanical Phenomena in Crystals. New York. Vol. 1, 2. 1964.
3. Kao K.C., Hwang W. Electrical Transport in Solids. With particular reference to organic semiconductors. Oxford-New York: Pergamon Press. Part 1, 2. 1981.
4. O'Dwyer I.I. The Theory of Electrical Conduction and Breakdown in Solid Dielectrics. Oxford: Clarendon Press. 1973.
5. Mason W.P. Crystal Physics of Interaction Processes. New York-London: Academic Press. 1966.
6. Horowitz P., Hill W. The Art of Electronics. Cambridge, New York: Cambridge University Press. 1998.
7. Mock M.S. Analysis of Mathematical Models of Semiconductor Devices. Dublin: Boole Press. 1983.
8. Uchino K. PZ Actuators and Ultrasonic Motors. Dordrecht, Netherlands: Kluwer Acad. Publ. 1996.
9. Cho Y.S., Yoon K.H. Dielectric Ceramics. In: Handbook of Advanced Electronic and Photonic Materials and Devices. "Ferroelectrics and Dielectrics". Ed. H.S.Nalwa. San Diego: Academ. Press. 2001.
10. Parton V.Z., Kudrjavtsev B.A. Electromagnetoelasticity of Solids: PZ and Electrically Conductive Materials. New York: Gordon Breach. 1988.
11. Physics of Electronic Ceramics. Part A. Ed. L.L.Hench, D.B.Dove. New York: Marcel Dekker. 1971.
12. Hench L.L., West J.K. Principles of Electronic Ceramics. New York: Wiley. 1990.
13. Kossov G. The effects of backing and matching on the performance of PZ ceramic transducers. IEEE Trans. Sonics. Ultrasonics. 1966. Vol.SU-13. No.1, p.20-30.
14. Drumheller D.S., Kalnins A. Dynamical shell theory for ferroelectric ceramics. JASA. 1970. Vol.65. No.3, p.1343-1353.

15. Holland R., Eer Nisse E. Design of Resonant PZ Devices. Cambridge, Massachusetts: MTI Press. 1968.

16. Mindlin R.D. Polarization Gradient in Elastic Dielectrics. Wien-New York: Udine CISM Sprunger. 1972.

17. R.D.Mindlin and Applied Mechanics. A Coll. Studies in Development of Applied Mechanics. Ed. G.Herrmann. New York: Pergamon. 1974.

18. Varadan V.K., Vinny K.J., Jose K.A. RF MEMS and their Applications. Atrium, South Gate, Chichester: J.Wiley & Sons. 2003.

19. Defects in SiO_2 and Related Dielectrics: Science and Techology. Ed. G.Pacchioni, L.Skuja, D.L.Griscom. Dordrecht-Boston-London: Kluwer Acad. Publ. Ser.II. Math. Phys. Chemistry. 2000.

20. Qin Q.H. Fracture Mechanics of Piezoelectric Materials. Southampton: WIT Press. 2001.

21. McFee J.H. Transmission and amplification of acoustic waves in PZ semiconductors. In: "Physical Acoustics". Part A. New York. P. 1-47.

22. Fantini E. et al. Failure mechanisms in compound semiconductor electron devices. In: Handbook of Advanced Electronic and Photonic Materials and Devices. Vol.2 "Semiconductor Devices". Ed. H.S.Nalwa. San Diego: Academ. Press. 2001. P.155-170.

23. Zelenka J. Piezoelectric Resonators and their Applications. Prague: Akademia. 1986.

24. Gautschi G. Piezoelectric Sensors: Force. Strain. Pressure. Acceleration and Acoustic Emission Sensors. Materials and Amplifiers. Berlin: Springer. 2002.

25. Khlifi Y. et al. Fowler-Nordheim current modeling of metal/ultra-thin oxide/semiconductor structures in the inversion node defects characterization. The European Phys. Journ. EPJ Appl. Phys. 2004. Vol.28. No.1, p.27-41.

26. Xiaofan Li, Daniels Chr. and Steinetz B.M. Optimized shapes of oscillating resonators for generating high-amplitude waves. JASA. 2004. Vol.116. No.5, p.2814-2821.

27. Paradies K. et al. Shape control of an adaptive mirror at different angles of inclination. J. of Intelligent. Mater. Systems and Struct. 1996. Vol.7. No.2, p.203-210.

28. Bruch J.C. et al. Optimal piezoactuator locations/length and applied voltage for shape control of beams. Smart Materials and Structures. 2000. Vol.9. No.2, p.205-211.

29. Irschik H. A review on static and dynamic shape control of structures by PZ actuation. Engineering Structures. 2002. Vol.24. No.1, p.6381-6403.

30. Pan E. Exact solution for simply supported and multilayered magneto-electro-elastic plates. J. Appl. Mechanics. 2001. Vol.68, p.608-618.

31. Pan E., Heyliger P.R. Free vibrations of simply supported and multilayer magneto-electro-elastic plates. J. Sound and Vibration. 2002. Vol.252, p.429-442.

32. Heyliger P.R. Static fields in magneto-electro-elastic laminates. AIAA Journ. 2004. Vol.42. No.7, p.1435-1444.

33. Mallik N., Ray M.C. Effective coefficients of PZ fiber-reinforced composites.

34. Lesiere G.A. Vibration damping and control using shunted PZ materials. Shock and Vibration Digest. 1998. Vol.30. No.3, p.187-195.

35. Ray M.C. Optimal control of thin laminated shells using PZ sensor and actuator layers. AIAA Journ. 2003. Vol.41. No.6, p.1151-1157.

36. Imry Y. Introduction in Mesoscopic Physics. Oxford: University Press. 2002.

37. Philips Data Handbook: "Components and Materials". Eindhoven: Philips. B. C9: Piezoelectric quartz devices. 1986. VII.

38. Evaluation of Materials and Structures by Quantitative Ultrasonics. Ed. J.D.Achenbach. Wien-New York: Springer. 1998.

39. Sasaki Y. et al. Small multilayer PZ transformers with high power density – characteristics of second and third-mode Rosen-type transformers. Jpn. J. of Appl. Phys. Part 1. Regular Papers, Short Notes and Review Papers. 1999. Vol.38. No.9B, p.5598-5602.

40. Tanuma Ch. A parallel-bimorph-type PZ actuator for high-resolution imager. Ibid., p.5603-5607.

41. Miyazawa O. et al. Investigation of ultrasonic motors using piezoceramics and a metal-composite-plate for watches. Ibid., p.5608-5611.

42. Abe H. et al. Trapped energy gyroscopes using thickness shear vibrations in a partially polarized PZ ceramic plate. Jpn. J. of Appl. Phys. Part 1. 1998. Vol.37. No.9B, p.5345-5348.

43. Ched-Ho Yun et al. A high power ultrasonic linear motor using a longitudinal and bending hybrid bolt-clamped Langevin type transducer. Jpn. J. of Appl. Phys. Part 1. 2001. Vol.40. No.5B, p.3773-3776.

44. Bhandari R., Miyazaki Y. High-speed non-reciprocal optical switching using phase shift induced by magnetostatic surface waves. Ibid., p.3768-3772.

45. Kapsh R.P., Kantz H., Hegger R., Diestelhorst M. Determination of the dynamical properties of ferroelectrics using nonlinear time series analysis. Int. J. Bifurcation and Chaos. 2001. Vol.11. No.3, p.1019-1034.

46. Erickson R.W. Fundamentals of Power Electronics. New York: Chapmen and Hall. 1997.

47. Rutledge D.B. The Electronics of Radio. Cambridge University Press. 1999.

48. Clark J.J., Zuille A.L. Data Fusion for Sensory Information Processing Systems. Boston-Dordrecht-London: Kluwer Academ. Publ. 1990.

49. Ibrahim K.F. Electronic Systems and Techniques. London: Longman Group UK. Sci. Techn. 1994.

50. Jones M.H. A Practical Introduction to Electronic Circuits. Cambridge University Press. 1995.

51. Rangayan M., Dhawan A.P., Gordon P. Algorithms for limited-view computed tomography: an annotated bibliography and challenge. Appl. Optics. 1985. Vol.24. No.23, p.4000-4012.

52. Natterer F. The mathematics of computerized tomography. New York: J.Wiley & Sons. 1986.

53. Streibe N. Three-dimensional imaging by a microscope. J.Opt.Soc.Amer.A. 1985. Vol.2. No.2, p.121-127.

54. Handbook of Pattern Recognition and Image Processing. Ed. T.Young. K.-S.Fu. New York: Academ. Pr. 1986.

55. Jayant N. Signal Compression: Coding of Speech, Audio, Text, Image and Video. Singapore: World Scientific. 1997.

56. Pennebaker W.B., Mitchell J.L. JPEG Still Image Data Compression Standard. New York: Van Nostrand Rienhold. 1993.

57. Algazi V.R., Avadhanam N. and Estes R.R. Quality measurement and use pre-processing in image compression. Signal Process. 1998. Vol.70, p.215-229.

58. Xiang S., Ni G. The Principle of the Photoelectric Imaging Devices. Beijing, China: Nat. Defense Industry Press. 1999.

59. Richardson R.L., Griffiths P.R. Generation of front-surface low-mass epoxy-composite mirrors by spin-casting. Optical Engineering. 2001. Vol.40. No.2, p.252-258.

60. Howes M.J., Morgan D.V. Eds. Charge-Coupled Devices and Systems. New York: Wiley. 1979.

61. Buchanan J.E. Bi CMOS/CMOS System Design. New York: McGraw-Hill, Inc. 1991.

62. Bregni S. Synchronization of Digital Telecommunications Networks. Chichester: J.Wiley & Sons, Inc. 2001.

63. Digital Optical Computing. Ed. R.A.Athale, W.A.Bellingham. USA: SPIE. 1990.

64. Optical Processing and Computing. Ed. H.Arsenault, T.Szoplin, B.Masukov. Boston-San Diego-New York: Academic Press. 1989.

65. Buca D.M. Fabrication and characterization of ultra-fact Si-based detectors for near-infrared wavelengths. Inaugural Diss. Forschungszentrum Jülich. 2002.

66. Opalenkov Y.V., Potapov A.A. Stochastic signals and Radon transformation when obtaining raster images by MW digital radio locator

with fractal information processing. Radiotekhnika i Elektronika. 2000. Vol.45. No.12, p.1447-1458.

67. Materials Science and Technology. A Comprehensive Treatment. Ed. R.W.Cahn, P.Haasen, E.J.Kramer. Vol.3A. Electronic and Magnetic Properties of Metals and Ceramics. Part I. Vol. Ed. K.H.-J.Buschow. Weinheim-New York-Basel-Cambridge. VCH. 1992.

68. Rogge T., Rummler Z., Schomburg W.K. Entwicklung eines piezogetrieben Mikroventils – von der Idee bis zur Vorserienfertigung. Karlsruhe: Forschungszentrum Karlsruhe GmbH. 2001, Wissensch. Berichte FZKA. P.6671.

69. Capasso F. Physics of Quantum Electron Devices. Ed. D.H.Auston. Berlin-Heidelberg: Springer-Verlag. 1990.

70. Gladkov S.O. Dielectric properties of porous media. New York: Springer. 2003.

71. Malygin G.A., Khusainov M.A. Stability analysis of the mechanical behaviour of a titanium nickelide at constrained shape memory (in Russian). Sov. Phys. – Techn. Phys. 2004. Vol.74, p.57-63.

72. Bogomol'nyi V.M. Design of piezoelectric actuators for measurement technology instruments. Measurement Techniques. 1998. Vol.41. No.7, p.654-660.

73. AIP Handbook of Condenser Microphones. Ed. G.S.K.Wong, T.F.W.Embleton. New York: AIP Press. 1995. (AIP – Amer. Inst. of Phys.)

74. Baum C.E. Target symmetry and the scattering dyadic. In: "Frontiers in Electromagnetic". Ed. D.H.Werner, R.Mittra. IEEE Press. 1999. Ch.4, p.204-236.

75. Baum C.E. Application of concepts of advanced mathematics and physics to the Maxwell equations. In: Ultra-Wideband, Short-Pulse Electromagnetics 5. Ed. P.D.Smith, S.R.Cloude. New York.

76. Gheewala T. The Josephson Technology. Proc. IEEE. 1982. Vol.70. No.1, p.26-34.

77. Shantz H.G. On the localization of electromagnetic energy. In: Ultra-Wideband, Short-Pulse Electromagnetics 5. Ed. P.D.Smith, S.R.Cloude. New York; Kluwer Academic/Plenum Publ. 2002. P.89-96.

78. Maxwell J.C. A Treatise on Electricity and Magnetism. Vol.II. Stanford. CA. Academic Repr. 1953, § 631.

79. Heavyside O. Electromagnetic Theory. Vol.1. New York: Chelsea Publ. Co. 1971.

80. Shanz H.G. The flow of electromagnetic energy around electric dipole. Am. J. Phys. 1995. Vol.63, p.513-520.

81. Kardo-Sysoev A.F. et al. Powerful sources of ultrawide band pulsed coherent signals. In: Ultra-Wideband, Short-Pulse Electromagnetics 5. Ed.

P.D.Smith, S.R.Cloude. New York; Kluwer Academic/Plenum Publ. 2002. P.335-342

82. Grekhov I.V., Efanov V.M., Kardo-Sysoev A.F. et al. Formation of high nanosecond voltage drop across semiconductor diode. Sov. Tech. Phys. Lett. 1983. Vol.9. No.4.

83. Kardo-Sysoev A.F. et al. Ultra-wide band solid state pulsed antenna array. Ibid., p.343-349.

84. Popov S. Integrated optoelectronics – the next technological revolution. In: Advanced Electronic Technologies and Systems Based on Low-Dimensional Quantum Devices. Ed. M.Balkanski, N.Andreev. Dordrech-Boston-London (printed in Netherland): Kluwer Academ. Publ. 1997. P.137-153.

85. Popov S. New integrated photoreceivers systems – charge-coupled devices. Ibid., p.175-187.

86. Shur M. GaAs devices and circuits. New York-London: Plenum Press. 1987.

87. The Infrared Handbook. Ed.: W.L.Wolfe, G.E.Zissis: Environmental Research Inst. of Michigan (ERIM). 1993. Vol.1-4.

88. Microcavities and Photonic Bandgaps: Physics and Applications. New York: Cluwer Academ. Publ. 1995.

89. Optical Processes in Microcavities. Ed. R.K.Chang, A.J.Campillo. Singapore: World Sci. 1996.

90. Hayashi C. Nonlinear Oscillations in Physical Systems. New York: McGraw-Hill. Repr. Prinston Univers. Press. 1984.

91. Cartwright M.L., Littlewood J.E. On nonlinear differential equation of the second order. J. London Math. Soc. 1945. Vol.20, p.180-189.

92. Feigenbaum M.J. Quantitative universality for a class of nonlinear transformations. J. Stat. Phys. 1978. Vol.19, p.25-52.

93. Feigenbaum M., Kadanoff L.P., Shenker S. Quasiperiodicity in dissipative systems: A renormalization group analysis. Physica. Vol.5D, p.370-386.

94. Flaherty J.E., Hoppensteadt F.C. Frequency entrainment of a forced van der Pol oscillator. Studies in Appl. Math. 1978. Vol.58, p.5-15.

95. Grebogi C. et al. Strange attractors that are not chaotic. Physica, 1984. Vol.13D, p.261-268.

96. Guckehheimer J., Holmes P. Nonlinear Oscillations, Dynamical Systems and Bifurcation of Vector Fields. New York: Springer-Verlag. 1983.

97. Meinardus G. , Nürnberg G. Delay Equation, Approximation and Application. Basel: Berkhauser. 1985.

98. Langford W.F. Numerical solution of bifurcation problems for ordinary differential equations. Numer.Math. 1997. Vol.28, p.171-190.

99. Levi M. Qualitative analysis of the periodically forced relaxation oscillations. Mem. Amer. Math. Soc. 1981. No.244.

100. May R.M. Simple mathematical models with very complicated dynamics. Nature. 1976. Vol.261, p.459-467.

101. Poincaré H., Ouevres I. Paris:Gaultmer-Villar. 1954.

102. Walther H.O. Dynamics of feedback systems with time lang. In: "Temporal Order". Ed. L.Rensing, N.I.Jaeger. Berlin-Heidelberg-London-New York-Tokio: Springer-Verlag. 1985. P.281-290.

103. Sri-Namachivaya N. et al. Non-standard reduction of noisy Duffing-van der Pol equations. Dynamical systems. An Int. Journ. 2001. Vol.16. No.3, p.223-246.

104. Buanomo A. The perturbation solution of van der Pol equation. SIAM. J. Appl. Math. 1998. Vol.59. No.1, p.156-171.

105. Anishechenko V.S., Neuman A.B. Stochastic synchronization. In: "Stochastic Dynamics. Lecture Notes in Physics. No.484. Ed. Th.Pöschel, L.Schimansky-Geier. Berlin:Springer-Verlag. 1997. P.155-166.

106. Baras F. Stochastic analysis of limit cycle behaviour. Ibid., p.167-178.

107. Marchesoni F. Stochastic resonance. Ibid., p.193-203.

108. Sadiki M.N.O. Elements of Electromagnetics. Fort Worth-Chicago-San Franscisco-Tokyo: Saunders College Publ. a Division of Holt, Rinehart and Winston, Inc. 1989.

109. Seeger K. Semiconductor Physics: Introduction 5 ed. Berlin: Springer. 1991. XIV. (Springer series: Solid State Sciences. No.40)

110. Olson B.W., Chen K. Microstructure defect detection using thermal response. Proc. SPIE. 2002. Vol.4735, p.714-725.

111. Sakagami T., Rubo S. Development of a new crack identification method based on singular current measurement using differential tomography. Proc. SPIE. 1999. Vol.3700, p.369-376.

112. Miy D.K. Mechatronics, Electromechanics. New York: Springer. 1993

113. Nayfeh A.H., Nemat-Nasser S. Electromagnetic-thermoelastic plane waves in solids with thermal relaxation. Trans. ASME. J. Appl. Mech. 1972. Vol.39, p.108-113

114. Yeh C. Dielectric waveguide theory. In: Recent Advances in Electromagnetic Theory. Ed. H.N.Kritikos, D.L.Jaggard. New York: Spinger-Verl. 1990, p.367-386

115. Goodwin G.C., Payne R.L. Dynamic system identification: Experiment, Design and Data Analysis. New York: Acad. Pr. 1977

116. Nayle A.H. Nonlinear oscillations. New York: Acad. Pr. 1979

117. Ragulskis K.M. Metal Vibromotors for precision microrobots. New York: Hemisphere. 1988

118.   Kochervinsky V.V. Piezoelectricity in crystallizing ferroelectric polymers as shown by PVDF and its copolymers. Kristallographiya. 2003. No.4, p.699-726 (in Russian)

119.   Bogomol'nyi V.M. To dynamical theory of electrothermal degradation and nondestructive testing of defects of metal-dielectric-metal (MDM) structures. SPIE Conf. Thermosense'99. Orlando, Florida. Apr. 1999. Proc. SPIE. 1999. Vol.3700, p.436-444

120.   Contact Adhesion and Rupture of Elastic Solids. Ed. D.Mangis. New York: Springer-Verlag. 2000

121.   Bogomol'nyi V.M. Design of PZ actuators for measurement technology instruments. Measurement Techniques. 1998. Vol.47. No.7, p.654-660

122.   Yarovikov V.I., Bazhenov A.A. Analysis of PZ ultrasonic receivers based on the equations of electroelasticity when the sensitive element is in state of bulk stress. Measurement Techniques. 1998. Vol.47. No.7, p.661-666